Cover:

The graphic representations are the flags of the five major carbon dioxide ($CO_2$) producing … polluting countries in the world that account for well over 50% of emissions.  Their emissions ranking for 2015 (slightly larger than those for 2014 shown on page 31):

| | |
|---|---|
| China | 28.03% |
| USA | 15.9% |
| India | 5.81% |
| Russia | 4.97% |
| Japan | 3.84% |
| **TOTAL** | **58.55%** |

https://www.statista.com/statistics/271748/the-largest-emitters-of-co2-in-the-world/

Retrieved 12/09/2016

1

# Aggressive Solutions
# for
# Climate Change

## by

## Duane R. Chartier

Foreword & Editorial Assistance:     Dr. Ann McMillan

Editorial Assistance:                Celeste Bryant

Editorial Assistance:                Dr. Ajeet Jon Saxon


Paperback    ISBN-10:0-9706115-2-8        ISBN-13:978-0-9706115-2-9

E-book       ISBN-10:0-9706115-1-X        ISBN-13:978-0-9706115-1-2

# CONTENTS

# THE AUTHOR

Dr. Duane R. Chartier was an inorganic chemist who had an early life crisis and decided to switch fields. He entered paintings conservation via degrees in art history and art conservation. Like most people, he has often wondered what the hell he is doing or supposed to be doing. *Ergo ... ipso facto*, this book.

His knowledge of fundamentals in materials science and computing led him to the Getty Conservation Institute in Los Angeles, CA, and thence to ICCROM (International Center for the Study of the Preservation and the Restoration of Cultural Property) in Rome, where he coordinated the scientific program. During this time and since then he has practiced as an independent consultant for projects ranging from computer system design at the Peggy Guggenheim Collection in Venice to wall paintings analysis and treatment consultation in the Residenz in Salzburg, Austria. During his career in art conservation he has moved very large sculptures and installed very large paintings. Some of his more curious dealings have been with the hand and footprints at Mann's Chinese Theatre in Hollywood, CA.

He is currently engaged in a number of activities that are personally interesting but likely won't earn a single dollar. What he does do to make money is provide the scientific support for authentication of works of art and fabrication of art supports. Most recently he has co-produced, with Roz Roembke, an online catalogue raisonné of the works of Wilson Hurley (www.wilson-hurley.com). In addition, he co-wrote, with the beautiful artist Alyson Souza (www.paintwood.com), *Materials Matter: An Artist's Book of Hours* which is an interesting collaboration between an artist and a materials scientist / art conservator. His most recent writing project is a collection of Aesopian short stories entitled *Notes from the Accretion Disk* (https://www.amazon.com/Notes-Accretion-Disk-Duane-Chartier/dp/0970611501).

He recently opened a new website (www.savant-tools.com) to sell honeycomb panels and cool tools. Many years ago, he made the bold mistake of starting the non-profit organization, The International Center for Art Intelligence, Inc. (ICAI, Inc.; www.authentica.org). His fondness for oxymorons is clearly illustrated in the name of the organization which, to date, has primarily been a magnet for lunatics.

All along the sinuous and rocky way he has always been a scientist at the core – just not a very patient or well behaved one. With the benefits of world experience and a number of added years he has finally realized that there is no point assuming that people will make the correct decisions without an enormous amount of encouragement, rather than direct criticism, which is also still sometimes necessary.

# FOREWORD BY DR. ANN MCMILLAN

Climate change is, perhaps, the biggest challenge to ever face us. It is here now, manifesting itself in changing sea levels, warming temperatures and shrinking ice, changing weather, increasing numbers and severity of extreme events, the arrival of new pests and diseases, and on and on. The world is rapidly becoming a different place than the one you and I grew up in. For our children, grandchildren, and posterity we need to manage our world in such a way as to meet this challenge. Time is running out.

The **science of climate is an exact science, based on observations and factual deductions from these**. On the other hand, opinions expressed about climate change fall on a distribution from the most ardent nay-sayers to the most devoted believers. As with many phenomena, this distribution is naturally bell shaped with the majority bunched in the middle not feeling too strongly either way (see figure). The "average" person is neatly in the middle and most people are in this group. The deniers are at one end of the spectrum and those who demand extreme actions to address the changing climate are at the other.

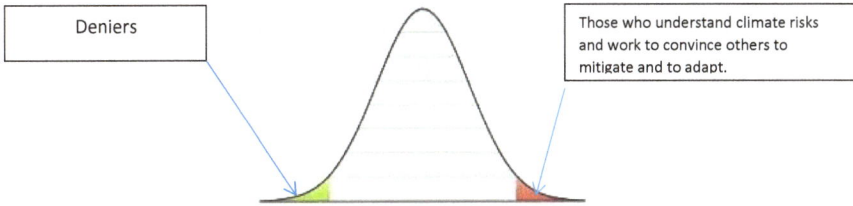

In Canada, under our previous leadership, government scientists were not allowed to publicly speak about issues such as this. Although our government has changed, it takes a while to recover from this situation, which seems to be becoming more common in other parts of the world. With the end of the spectrum which generally understands the science being effectively silenced, the distribution has inevitably shifted toward the nay-sayer end. This is unfortunate since it is important in this societal discussion to consider a spectrum of opinions which span the whole distribution in order to have an inclusive debate and develop an acceptable, workable path forward in time.

It is interesting that discussions about nuclear energy have also seen a similar distribution shift with the nuclear industry being nearly silent in terms of promoting the importance and value of nuclear energy.

Sadly, all involved in both issues seem to have lost their sense of humour along the way.

Dr. Duane Chartier is intent on re-establishing the voice of the previously silent end of the spectrum on the two extremely important issues of the changing climate and the role of nuclear energy. This book fills an important gap by opening the dialogue to include those who understand the science and are alarmed at the potential for changing climate to alter the world in ways that will be difficult or impossible for future generations to manage. The book looks not only at the well-accepted current status of the climate, but also at the probable futures facing us. It identifies energy as being one of the major dimensions of the challenge and goes even further to suggest "no regrets" solutions which will develop existing and new nuclear technologies to be safe, clean, and economically sound. All this and it does so with a light touch and humour ranging from silly to deepest black.

> *Dr. Ann McMillan – Ann McMillan has had a varied career as a scientist in the Canadian environmental industry and government including:*
>
> > *Fisheries and Oceans Canada - Special Advisor on Climate Change and Ocean Science*
> >
> > *Meteorological Service of Canada - Director Strategic Policy, specialist in Air Quality*
> >
> > *Ontario Hydro – Research Division, Section Head.*
>
> *She is presently semi-retired but is President of Storm Consulting. She is active in the Canadian Meteorological and Oceanographic Society (CMOS) and on the Board of Directors of the Canadian Climate Forum (CCF). Her background is in mathematics, computer science and mechanical engineering.*
>
> *In 2013, she co-edited a best selling technical book on the Canadian Climate (Air Quality Management: Canadian Perspectives on a Global Issue 2014th Edition - by Eric Taylor (Editor), Ann McMillan (Editor) https://www.amazon.com/Air-Quality-Management-Canadian-Perspectives/dp/9400775563/ref=sr_1_13?ie=UTF8&qid=1494967051&sr=8-13&keywords=ann+McMillan).*

# INTRODUCTION

*"You must be the change you wish to see in the world."*

Mahatma Gandhi (Indian Leader, October 2, 1869 - January 30, 1948)

A couple of years ago I had to face serious changes in my own life as we all must do from time to time. I had hidden behind my business and family in order to avoid the battles in the trenches that I had fought and lost years before. My son Raef inadvertently stumbled into an abyss when he asked me a question about statistics. A wound in my side seemed to open and I poured forth a story of woe that was one of the reasons I left a research job at Ontario Hydro Research in 1986.

I was editing a research publication for my boss (whose name appeared as editor) entitled <u>Research Review / Acid Rain (2)</u>. It was very early in the era of increasing environmental consciousness and we were an Atmospheric Research Unit. I took a paper, written by an engineer in the power generation division, to my boss and told him that we should not publish it. Basically, it said that given the precipitation chemistry data for eastern North America there was no statistically significant trend in the chemistry and acid rain was not apparent. I found this conclusion impossible to fathom as I did the specific statistical approach that was taken. At best, it was an overly simplistic and naïve approach to looking at data – not one based on science and deep insight. My boss reviewed the paper and told me that my concerns were overblown and that we should publish it because he saw no problem using a simple linear regression approach for looking at data. At great length and considerable elevation of my blood pressure, I explained why that approach could not possibly yield any useful analysis. It was to no avail. In typical scientific fashion, I just went above him to the director of research and said that we would look like asses if we published that paper. He was sympathetic but really did not want to create a "staff problem" by countermanding one of the few engineers in charge in a scientific section. The stunning power of bureaucracy finally became clear to the organizationally handicapped - me.

Much to the director's credit, he came up with a "solution" ... I could publish my own paper, back-to-back, with the other and "let people decide". To me, this was like letting people in the flat earth society decide on maritime navigation[1].

Of course, I was angry by this point. What about improper did my management not understand? Had I failed to communicate in the clearest and most concise manner? Was the proper use of statistics not a significant issue in a scientific organization? Was I about to lose my mind?

---

[1] I really think that this stamp (*the **A**uthor's **P**ersonal **V**irtual **G**raffiti **P**rogram (APVGP))* is the most efficient way to deal with inane arguments and obfuscation. I have tried to use it sparingly but, given the level of nonsense in climate discussions and political debate, it is incredibly difficult to use it sparingly.

I wrote the paper and it was published: D.R. Chartier. "Historical trends in precipitation chemistry in eastern North America, Part 2." Research Review / Acid Rain (2). Toronto: Ontario Hydro Research Division, May 1981. pp. 43 - 47. As far as I can tell my boss was the only one who actually read it and was very displeased with "my attitude" (whatever that meant). So, using the identical data, Part 1 (the other author) said that acid rain was not a problem in eastern North America and Part 2 (me) said that it was a significant problem. Seeing them both together made my existence totally questionable. It is, or it is not! – Science or not science! Much to my surprise and delight I was very quickly transferred from his unit to another where I could become someone else's "attitude problem".

My poor son had to listen to my heartfelt story about the right and true use of statistics, the evils of bureaucracy, the rigidity of some engineers and the general insanity of the world. The real point that I missed at the time and in my story to my son was that **I was in the wrong place, at the wrong time, saying the right things but in the wrong way**. People don't like to hear that they are wrong. No one really wants to feel stupid, slow or dense. So, in the words of an ex-wife I should work on my "style" – roughly meaning, I have to make people feel good about not doing the right thing, having indefensible positions, or even a modicum of intelligence. I guess the answer to the question, "Does my dress make me look fat?" is definitively NOT, "No, it's not the dress." Hopefully, thirty or more years after leaving science and a couple of years after becoming divorced a second time (isn't it good that I didn't say becoming a free man?), I can get back on track and try to do the right thing with no excuses.

Being forced to accept these things about myself in the context of others has taught me the following two things, which mean more to me later in life than they ever meant before.

> *"Progress is impossible without change, and those who cannot change their minds cannot change anything."*

> George Bernard Shaw (Irish Playwright and Political Activist, July 26, 1856 - November 2, 1950)

> *"Never believe that a few caring people can't change the world. For, indeed, that's all who ever have."*

> Margaret Mead (American Cultural Anthropologist, December 16, 1901 - November 15, 1978)

Change is what we must all do!  The three quotations invoking change are reminders for all of us to not stop saying what we think is right just because it may be unpopular or uncomfortable.  Indeed, we must change how we see the world, stop making poor excuses for not doing anything as being "human nature", waiting for someone else to do it first, waiting for approval, trusting politicians to do the thinking for us.  If we are to beat back the potential disaster of Climate Change we must change our thinking.  The purpose of the book is not to simply reiterate what is already known but to try to inject some small, hopefully viral ideas, into the often nonsensical discussion of energy policy.  We need nuclear power, especially fusion; we need a global approach to planetary management; we need solutions not rhetoric and pointless debate.

*Duane R. Chartier – Recovering Scientist and perhaps, New-found Teetering Optimist*

# GLOBAL WARMING IS REAL

Despite naysayers and intellectual ostriches, global warming is real and the climate crisis is imminent. Perhaps the word "crisis" is too extreme for many people and they would prefer a more comfortable adjective. However, there are some fundamental historical trends that we should understand and incorporate into our overall approach to climate and climate change.

Starting with basics is always difficult when nature is not "clean" and tractable like simple equations. As stated directly in the introduction, it is critical not only to have a statistical view of nature but to have "the right statistical view" of nature. The intrinsic problem in discussing climate change is that we must encompass its statistical nature. It is often difficult to convince somebody that there is global warming when "Winnipeg experiences second snowiest December in city's history" in 2016.[2] People usually focus on the local ... not the global.

Humans are notoriously bad at integrating statistics within a temporal context. There would be no casinos in Las Vegas if people didn't have the strange notion that they can win money. YES, it is individually possible to win money. NO, it is collectively, statistically, impossible to beat the casinos in the long term. The house has better odds at every game and will win ... eventually ... in time.

"According to an ongoing temperature analysis conducted by scientists at NASA's Goddard Institute for Space Studies (GISS), the average global temperature on Earth has increased by about 0.8° Celsius (1.4° Fahrenheit) since 1880. Two-thirds of the warming has occurred since 1975, at a rate of roughly 0.15-0.20°C per decade."[3]

So, the measured and confirmed global warming (mostly since 1975) is 0.015 – 0.020°C per year. This is averaged for the entire planet. This seems very small so even if we accept this, how do we translate the effects of these increases into phenomena that average people might relate to other than what the particular temperature is outside their house today? Some attempt will be made over the next several pages to make the numbers more meaningful but we must look at the numbers first and realize that there is no conspiracy here or even questionable measurements as will be discussed later.

The following table indicates only one of many sources confirming global temperature change.

---

[2] "Winnipeg experiences second snowiest December in city's history". Emad Agahi, CTV Winnipeg. Published Friday, December 30, 2016 7:44PM CST. http://winnipeg.ctvnews.ca/winnipeg-experiences-second-snowiest-december-in-city-s-history-1.3222919. Retrieved 02/21/2017.

[3] http://earthobservatory.nasa.gov/Features/WorldOfChange/decadaltemp.php. Retrieved 11/09/2016.

## Global Land-Ocean Temperature Index (°C) (Anomaly with Base: 1951-1980)

| Year | Annual Mean | 5-year Mean | Year | Annual Mean | 5-year Mean | Year | Annual Mean | 5-year Mean | Year | Annual Mean | 5-year Mean |
|------|------|------|------|------|------|------|------|------|------|------|------|
| 1880 | -0.19 | * | 1916 | -0.33 | -0.24 | 1952 | 0.01 | -0.05 | 1988 | 0.40 | 0.33 |
| 1881 | -0.10 | * | 1917 | -0.39 | -0.26 | 1953 | 0.08 | -0.05 | 1989 | 0.29 | 0.38 |
| 1882 | -0.08 | -0.16 | 1918 | -0.25 | -0.29 | 1954 | -0.12 | -0.07 | 1990 | 0.44 | 0.35 |
| 1883 | -0.19 | -0.19 | 1919 | -0.22 | -0.26 | 1955 | -0.14 | -0.07 | 1991 | 0.42 | 0.32 |
| 1884 | -0.26 | -0.22 | 1920 | -0.25 | -0.24 | 1956 | -0.19 | -0.07 | 1992 | 0.23 | 0.33 |
| 1885 | -0.30 | -0.27 | 1921 | -0.20 | -0.23 | 1957 | 0.04 | -0.04 | 1993 | 0.24 | 0.33 |
| 1886 | -0.29 | -0.27 | 1922 | -0.26 | -0.24 | 1958 | 0.06 | -0.02 | 1994 | 0.32 | 0.32 |
| 1887 | -0.32 | -0.24 | 1923 | -0.23 | -0.23 | 1959 | 0.03 | 0.03 | 1995 | 0.46 | 0.37 |
| 1888 | -0.19 | -0.25 | 1924 | -0.27 | -0.21 | 1960 | -0.03 | 0.03 | 1996 | 0.34 | 0.44 |
| 1889 | -0.10 | -0.24 | 1925 | -0.19 | -0.19 | 1961 | 0.05 | 0.03 | 1997 | 0.47 | 0.47 |
| 1890 | -0.35 | -0.22 | 1926 | -0.09 | -0.19 | 1962 | 0.02 | -0.02 | 1998 | 0.63 | 0.46 |
| 1891 | -0.23 | -0.24 | 1927 | -0.20 | -0.20 | 1963 | 0.06 | -0.04 | 1999 | 0.42 | 0.50 |
| 1892 | -0.25 | -0.28 | 1928 | -0.21 | -0.19 | 1964 | -0.21 | -0.06 | 2000 | 0.42 | 0.53 |
| 1893 | -0.28 | -0.25 | 1929 | -0.35 | -0.19 | 1965 | -0.10 | -0.07 | 2001 | 0.54 | 0.53 |
| 1894 | -0.29 | -0.23 | 1930 | -0.13 | -0.18 | 1966 | -0.05 | -0.09 | 2002 | 0.63 | 0.55 |
| 1895 | -0.20 | -0.20 | 1931 | -0.08 | -0.20 | 1967 | -0.03 | -0.04 | 2003 | 0.62 | 0.60 |
| 1896 | -0.14 | -0.20 | 1932 | -0.15 | -0.15 | 1968 | -0.07 | -0.01 | 2004 | 0.54 | 0.62 |
| 1897 | -0.10 | -0.17 | 1933 | -0.27 | -0.16 | 1969 | 0.06 | -0.02 | 2005 | 0.69 | 0.63 |
| 1898 | -0.27 | -0.15 | 1934 | -0.12 | -0.17 | 1970 | 0.03 | -0.01 | 2006 | 0.63 | 0.61 |
| 1899 | -0.15 | -0.15 | 1935 | -0.18 | -0.14 | 1971 | -0.09 | 0.03 | 2007 | 0.66 | 0.63 |
| 1900 | -0.08 | -0.18 | 1936 | -0.14 | -0.09 | 1972 | 0.01 | 0.00 | 2008 | 0.53 | 0.64 |
| 1901 | -0.14 | -0.20 | 1937 | -0.01 | -0.07 | 1973 | 0.15 | -0.00 | 2009 | 0.64 | 0.63 |
| 1902 | -0.27 | -0.25 | 1938 | -0.01 | -0.02 | 1974 | -0.08 | -0.01 | 2010 | 0.72 | 0.62 |
| 1903 | -0.35 | -0.29 | 1939 | -0.02 | 0.04 | 1975 | -0.01 | 0.03 | 2011 | 0.60 | 0.65 |
| 1904 | -0.43 | -0.31 | 1940 | 0.09 | 0.06 | 1976 | -0.11 | 0.01 | 2012 | 0.63 | 0.67 |
| 1905 | -0.27 | -0.33 | 1941 | 0.13 | 0.09 | 1977 | 0.18 | 0.06 | 2013 | 0.65 | 0.70 |
| 1906 | -0.21 | -0.35 | 1942 | 0.10 | 0.15 | 1978 | 0.07 | 0.12 | 2014 | 0.74 | * |
| 1907 | -0.39 | -0.35 | 1943 | 0.14 | 0.15 | 1979 | 0.17 | 0.20 | 2015 | 0.87 | * |
| 1908 | -0.42 | -0.38 | 1944 | 0.26 | 0.12 | 1980 | 0.28 | 0.19 | | | |
| 1909 | -0.46 | -0.43 | 1945 | 0.13 | 0.09 | 1981 | 0.33 | 0.24 | | | |
| 1910 | -0.42 | -0.42 | 1946 | -0.03 | 0.05 | 1982 | 0.13 | 0.24 | | | |
| 1911 | -0.44 | -0.40 | 1947 | -0.04 | -0.02 | 1983 | 0.31 | 0.21 | | | |
| 1912 | -0.34 | -0.34 | 1948 | -0.09 | -0.08 | 1984 | 0.16 | 0.18 | | | |
| 1913 | -0.33 | -0.27 | 1949 | -0.09 | -0.09 | 1985 | 0.12 | 0.22 | | | |
| 1914 | -0.15 | -0.25 | 1950 | -0.18 | -0.08 | 1986 | 0.19 | 0.24 | | | |
| 1915 | -0.10 | -0.26 | 1951 | -0.07 | -0.05 | 1987 | 0.34 | 0.27 | | | |

http://climate.nasa.gov/system/internal_resources/details/original/647_Global_Temperature_Data_File.txt   Retrieved 11/09/2016.

*Since 1977 there is no negative annual mean temperature recorded!*

The data above is more accessible or comprehensible as a graph (presented below):

*"This graph illustrates the change in global surface temperature relative to 1951-1980 average temperatures. The 10 warmest years in the 134-year record all have occurred since 2000, with the exception of 1998. The year 2015*

*ranks as the warmest on record. (Source: NASA/GISS). This research is broadly consistent with similar constructions prepared by the Climatic Research Unit and the National Oceanic and Atmospheric Administration."*

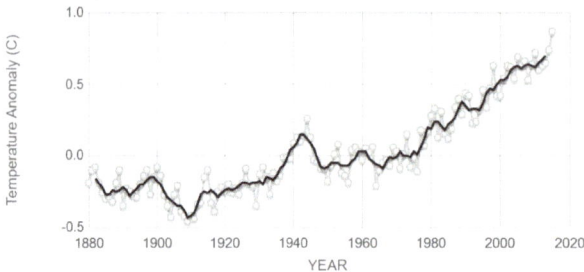

http://climate.nasa.gov/vital-signs/global-temperature/

We can only truly understand global warming by taking a global and necessarily, statistical approach to climate and this data is an honest starting point.

Although websites change at the blink of an eye it is both interesting and alarming to note the bottom page banner on the NASA website (http://climate.nasa.gov/ Retrieved 03/10/2017). The four bullets are captured in the screenshot below:

| CARBON DIOXIDE | GLOBAL TEMPERATURE | ARCTIC ICE MINIMUM | LAND ICE |
|---|---|---|---|
| ↑ 405.92 parts per million | ↑ 1.7 °F since 1880 | ↓ 13.3 percent per decade | ↓ 281.0 Gigatonnes per year |

The first bullet is very significant in that carbon dioxide ($CO_2$) is the best single tracker of human contribution to global warming. For a minimum of 400,000 years $CO_2$ had never exceeded 300 ppm in the atmosphere until 1950. This will be explored in more detail in the next section. So, aside from the Ross Ice Shelf disappearing, hurricane strengths intensifying, accelerating desertification, and many recent record setting years for temperature increase, it is difficult to know what further evidence of a problem is required.

***A personal challenge to all climate change naysayers!*** – I will match the funds you spend on any property presently (and verifiably) at sea level today (June 05, 2017) if, within ten years (and without seawalls, and any other mitigating protections) it has not had any measurable encroachment by the rising water levels. Of course, this bet is conditional in that there are elements of natural tectonic rise and convective geological behavior that may alter the perceived rise locally but they can be subtracted objectively and don't count in the bet. If there is encroachment you agree to pay me the equivalent match of funds. All agreements must be made in writing within 1 year of this limited time offer! If the book doesn't sell I could certainly retire on this wager because there is virtually no way I will lose!

# FOSSIL FUELS & GLOBAL WARMING

*We simply must balance our demand for energy with our rapidly shrinking resources. By acting now we can control our future instead of letting the future control us."*

Jimmy Carter (39th President of the United States - January 20, 1977 – January 20, 1981 – from a televised speech April 18, 1977.)

The "energy crisis" of the late 1970's was artificial but, none-the-less, a powerful warning about energy use and international politics. The US Department of Energy was formed under the leadership of President Jimmy Carter in order to better understand and control the production, distribution and use of energy.

The production of energy in the world is, at present, seriously tied to climate and climate change. The simple and undeniable reason is the burning of fossil fuels.

Since our earliest ancestors "discovered", or perhaps better put, "mastered" fire, we have been on a collision course with chemical and environmental reality. **The burning of any organic material leads to the production of carbon dioxide ($CO_2$)**. Whether we casually sit beside the fireplace and put up our feet on the ottoman and relax with a good book and a glass of wine or we consume tons of coal or oil per hour to send the container ship from Singapore to Los Angeles we are producing $CO_2$. Lighting the candles on the birthday cake is also a willful and vile act of $CO_2$ production.

The combustion (oxidation) of any hydrocarbon will result in the production of $CO_2$. **We can't avoid this consequence** because the very basic chemistry / physics of the reaction is:

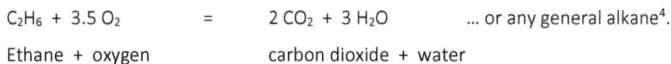

$C_2H_6$ + 3.5 $O_2$       =       2 $CO_2$ + 3 $H_2O$       ... or any general alkane[4].

Ethane + oxygen              carbon dioxide + water

It is important to include the combustion of ethanol in this discussion because it is a major additive to many gasolines in the USA[5] and is used extensively in countries like Brazil.

---

[4] Some of us must summon up basic organic chemistry – Alkanes are single-bonded, completely saturated hydrocarbons (ethane, propane, octane), alkenes (ethane, pentene, etc.) have at least one double bond and are unsaturated as are alkynes (acetylene) that have a triple bond.

[5] "In 2014, about 13 billion gallons of fuel ethanol were added to the motor gasoline produced in the United States. Fuel ethanol accounted for about 10% of the total volume of finished motor gasoline consumed." From the EIA website http://www.eia.gov/tools/faqs/faq.cfm?id=27&t=10.

$$2\ CH_3OH\ +\ 3\ O_2\qquad =\qquad 2\ CO_2\ +\ 4\ H_2O$$

$$\text{Ethanol}\ +\ \text{oxygen}\qquad =\qquad \text{carbon dioxide}\ +\ \text{water}$$

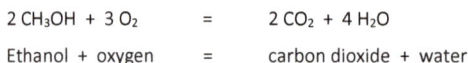

The "fossil fuels"[6] all originate with wood and are presented in order of historical use and energy yield:

| | Material | First Use | Specific Energy (MJ/kg)[7] |
|---|---|---|---|
| 0. | Wood ... | Pre-Neolithic | 17.1 |
| 1. | Coal | Neolithic[8] | 23.9 (bituminous) |
| 2. | Oil | 6th Century BCE[9] | 41.9 (crude) |
| 3. | Natural Gas | 500 BCE[10] | 47.2 |
| | *For comparison Uranium ... (more than 1 million times the energy density of anything above)* | | |
| | $^{235}U$ | Aug 6, 1945[11] | 79,390,000[12] |

The chemistry of the oxidation / combustion of wood and all fossil fuels is essentially the same for the sake of simple argument.

The problem with wood as a fuel is that it has not been broken down by the geological pressures and biological action to a purer hydrocarbon. Wood is primarily cellulose with a large amount of bound and unbound water. Anyone who has tried to start a fire with freshly cut wood knows what a failure that can be. Wood needs considerable drying before it can be burned. The equilibrium moisture content (EMC) of wood exposed to the atmosphere varies depending upon temperature, relative humidity and

---

[6] The **fossil fuels** - coal, oil, natural gas - are the plant and animal materials that were converted by burial, pressure, heat, biological action and time into useable fuel. Due to the much longer existence of plants and the much higher biomass, they contribute most of the starting material for any deposits of fossil fuels. The extremely rich reserves of coal in many parts of the world date to the "Carboniferous" period (from the end of the Devonian Period, about 358.9 ± 0.4 million years ago, to the beginning of the Permian Period, about 298.9 ± 0.15 million years ago). Oil and gas deposits are more associated with marine sediments formed 300-400 million years ago. The shallower deposits tend to be oil (less decomposed material) then natural gas and finally methane (completely broken down – the lowest numbered hydrocarbon – only 1 Carbon).

[7] Data abstracted from: C. Ronneau (2004), Energie, pollution de l'air et developpement durable, Louvain-la-Neuve: Presses Universitaires de Louvain.

[8] See "A Brief History of Coal" segment on the US Department of Energy website - http://www.fossil.energy.gov/education/energylessons/coal/coal_history.html. Retrieved 06/09/2015.

[9] 6th Century BC – "The army of Kir II, first shah of Achaemenid Empire (present - Iran), used Absheron oil in weapons of fire to invade castles and cities. (Note: much of the historic information relating to the early history in the Middle East has been provided by Mir-Yusif Mir-Babayev, Professor of Azerbaijan Technical University in Baku)." Retrieved 06/09/2015 from http://www.geohelp.net/world.html.

[10] The Chinese used bamboo pipelines to collect escaping natural gas and use it to boil water to create salt from seawater. Primary source not available, retrieved from http://naturalgas.org/overview/history/ 06/09/2015.

[11] "Little Boy" atomic bomb dropped on Hiroshima. Filling: Uranium-235, Filling weight: 140 lb (64 kg), Blast yield: 15 kilotons of TNT (63 TJ). See http://en.wikipedia.org/wiki/Little_Boy. Retrieved 06/09/2015.

[12] Datum abstracted from "Table 1. Energy densities of nuclear fuels." on http://www.whatisnuclear.com/physics/energy_density_of_nuclear.html. Retrieved 06/09/2015.

species but one can generally place it between a low of 9% to upwards of 20%.[13]  For this reason alone wood is not an efficient fuel and due to inefficient combustion, it tends to be a very dirty one.  Burning wood was fine for cave men but we have better alternatives.

The table below lists the relative amounts of carbon dioxide emitted when various fossil fuels are burned.

**Pounds of $CO_2$ emitted per million Btu of energy for various fuels:[14]**

| | |
|---|---|
| Coal (anthracite) | 228.6 |
| Coal (bituminous) | 205.7 |
| Coal (lignite) | 215.4 |
| Coal (subbituminous) | 214.3 |
| Diesel fuel & heating oil | 161.3 |
| Gasoline | 157.2 |
| Propane | 139.0 |
| Natural gas | 117.0 |

Natural gas produces the lowest amount of $CO_2$ per unit energy (about half as much as coal).  However, the global excess of $CO_2$ production is at the heart of global warming.  Regardless of how good natural gas is relative to other fossil fuels it is still NOT GOOD.  There are many peer-reviewed scientific articles on the correlation of $CO_2$ levels with global warming[15].

If we are experiencing significant changes in global climate how can we blame modern man for the changing global climate?[16]  The answer is actually somewhat simple but it comes in stages.

---

[13] William T. Simpson.  "Equilibrium Moisture Content of Wood in Outdoor Locations in the United States and Worldwide".  United States Department of Agriculture, Forest Service, Forest Products Laboratory Research Note FPL-RN-0268, 1998.

[14] Retrieved 06/09/2015 directly from the webpage of the US Energy Information Administration - http://www.eia.gov/tools/faqs/faq.cfm?id=73&t=11.

[15] "Global warming preceded by increasing carbon dioxide concentrations during the last deglaciation".  Jeremy D. Shakun, Peter U. Clark, Feng He, Shaun A. Marcott, Alan C. Mix, Zhengyu Liu, Bette Otto-Bliesner, Andreas Schmittner & Edouard Bard.  *Nature* 484, 49–54 (05 April 2012).  This is just one representative source that incontrovertibly shows the correlation.

[16] This is a bit of a catty comeback for those who would insist that the recent trends and contribution of the past global climate change are the same or not correlated.

# ELEMENTS OF CLIMATE CHANGE

Regardless of the unreferenced, improperly referenced or falsely reported "facts" found all over the Internet there are verifiable facts and sources of information regarding climate change. There is, MOST CERTAINLY, GLOBAL WARMING! The following materials attempt to present and explain, as simply as possible, the evidence for global warming. In part, the explanations will hopefully help in understanding why there may be some confusion.

One of the major problems in dealing with any global phenomenon is that its overall characterization is statistical but all climate manifestations are inherently local in the perception of people on the ground and those not involved in monitoring and modelling. Just because Boston had a very cold winter does not mean that there is no global warming!

What makes the climate change?

1.  Astronomical Changes – changes in the earth's position about the sun, changes in earth tilt, changes in the sun's actual output.

2.  Atmospheric Changes – Heat retention due to atmospheric gases, water vapor and the "greenhouse gases" - carbon dioxide ($CO_2$), methane, and a few others. Changes in solar reflectivity - volcanic dust, polar ice caps.

3.  Tectonic Changes - Shifting continents (continental drift) cause changes in circulatory patterns of ocean currents. There is strong paleoclimatic evidence[17] that when there is a large land mass at one of the Earth's poles, there tends to be a large cooling trend (ice age). This is, of course, strongly linked to the tilt of the earth and the relative reflectance of materials. Undersea ridge activity: "Sea floor spreading", normal plate tectonics, can cause variations in ocean displacement. This tends to be a very small effect relative to the others.

    The tectonic changes are not really an issue within the present focus on global warming for two very basic reasons. They occur over very long periods of time (millions of years) and they would not manifest themselves as large changes over the short time period phenomena that we need to consider. Therefore, we can very credibly disregard them for this discussion.

## ASTRONOMICAL CHANGES

The astronomical changes in the earth's climate are cyclic and, within the last 5-10 million years, are reversible or follow a flattened curve. Regardless of this fact we must know them and their characteristics very well to be able to dismiss them from the climate data (make calibrated corrections) so that the actual changes and sources can be tracked and understood.

---

[17] Paleoclimatology is the discipline that studies historical changes in the earth's climate by a variety of methods to producing overlapping and auto-verifying data. Methods include the study of polar ice cores, lake and ocean sediment cores and earth cores.

The most common astronomical changes that everyone recognizes are the four seasons. Most of us have learned in grade school that the rotation axis of the earth has an approximately 23° tilt relative to earth's orbital plane about the sun – this is called obliquity[18] and even this changes slightly over long periods of time like a slightly wobbling top (precession). The very large temperature differences of the season tend to wash out the smaller changes in average global temperature over the very long time periods that the wobbling or precession takes place. The cycle of the precession is about 26,000 years[19].

Besides the very dramatic seasonal changes in climate the next most frequent astronomical climate change is caused by variation in the sun itself. The 11-year cyclic variation of the sun has been known at least since the 4th century BCE[20] by Chinese astronomers. Shi Shen made a star map and star catalogue as well as wrote a book on astronomy. In 364 BCE, Gan De made the first recorded observation of sunspots, and the moons of Jupiter, and both astronomers made accurate observations of the five major planets[21].

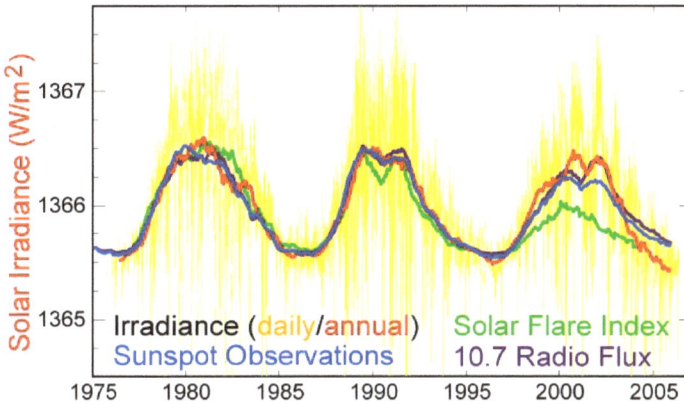

## Solar Cycle Variations

*ACTIVITY CYCLES 21, 22 AND 23 SEEN IN SUNSPOT NUMBER INDEX, TSI, 10.7CM RADIO FLUX, AND FLARE INDEX. THE VERTICAL SCALES FOR EACH QUANTITY HAVE BEEN ADJUSTED TO PERMIT OVERPLOTTING ON THE SAME VERTICAL AXIS AS TSI. TEMPORAL VARIATIONS OF ALL QUANTITIES ARE TIGHTLY LOCKED IN PHASE, BUT THE DEGREE OF CORRELATION IN AMPLITUDES IS VARIABLE TO SOME DEGREE. SOURCE: THIS IMAGE WAS CREATED BY ROBERT A. ROHDE FROM THEPUBLISHED DATA LISTED BELOW AND REPLACES AN IMAGE CREATED BY WILLIAM M. CONNOLLEY. IT IS PART OF THE GLOBAL WARMING ART PROJECT.*

---

[18] This obliquity changes over time. In the last 5 million years the variation has only been from about 22.0° to 24.5°. Source: Berger, A.L. (1976). "Obliquity and Precession for the Last 5000000 Years". *Astronomy and Astrophysics* **51**: 127–135.

[19] http://www.astro.cornell.edu/academics/courses/astro201/earth_precess.htm. Retrieved 06/08/2015.

[20] BCE = **B**efore the **C**hristian **E**ra. BC (**B**efore **C**hrist) is now less used.

[21] http://nrich.maths.org/6843. "Early Astronomy and the Beginnings of a Mathematical Science". NRICH (University of Cambridge). 2007. Retrieved 06/07/2015.

The variation in solar activity follows an 11-year cycle of solar maxima (highest energy output and sunspot production) and solar minima[22]. This is a result of intrinsic stellar seismology or, more accurately, it is the result of "a hydromagnetic dynamo process, driven by the inductive action of internal solar flows"[23]. This explanation is more than a bit obscure but the sun's burning of heavier elements and the relative amounts of hydrogen and helium in the core causes periodic internal changes that are re-equilibrated in the periods of higher activity.

These variations in solar activity have been observed for hundreds of years via sunspots and aurorae. The cycles are very regular and can be seen in the diagram. It is clear that although there is variation one can simply draw a horizontal line (no slope) representing the average solar activity. Therefore, although the sun does get "hotter" every 11 years it also gets about equally colder. **The net effect on earth climate is effectively zero but if one is looking at any more temporally limited curve of earth temperatures it is a factor that must be accounted for**.

An even longer period astronomical effect is caused by periodic changes in the actual shape of earth's orbit around the sun. These are changes in the eccentricity[24] of earth's orbit occur over a 100,000-year period (see the illustration on the next page).

At present the earth's orbit is very close to circular[25], meaning that the seasonal variations are the smallest we would experience in a yearly orbit around the sun. These 100,000 year cycles are caused by the changing alignments of the planets as they orbit the sun. The gas giants (Jupiter, Saturn, Neptune and Uranus) have a very large influence on the distortion of earth's orbit due to their combined masses and a simple geometric / temporal model would show that the maximal distortion of earth's orbit would occur when they are all aligned (in conjunction). Of course, this is relatively rare due to their very different orbital periods.

---

[22] My now 24-year-old daughter will never forgive me for encouraging a trip to Winnipeg, Manitoba from Los Angeles in the dead of winter, January 2006. When we landed the temperature was -49 C. The purpose of the trip was to see relatives and observe the Aurora Borealis for her homeschooling project. As educational consultant, I did not tell her that it was solar minimum – she was supposed to know that or, at least, research it! Needless-to-say, I was somehow responsible for absolutely no sunspots, therefore, no aurora and for the "bloody cold climate".

[23] "Solar Cycle", Wikipedia. Retrieved 06/07/2015.

[24] By definition, a circle has an eccentricity of 0 and as it becomes an ellipse the eccentricity increases.

[25] Currently, the Earth is experiencing a period of low eccentricity. "The difference in the Earth's distance from the Sun between perihelion and aphelion (which is only about 3%) is responsible for approximately a 7% variation in the amount of solar energy received at the top of the atmosphere. When the difference in this distance is at its maximum (9%), the difference in solar energy received is about 30%." Source: http://www.eoearth.org/view/article/51cbee337896bb431f696153/. Retrieved 06/08/2015.

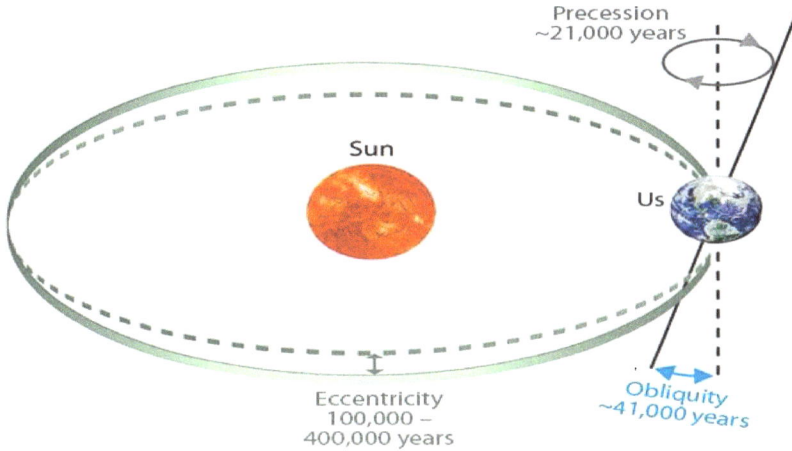

*THIS IMAGE WAS RETRIEVED (06/08/2015) FROM THE "MILANKOVITCH ORBITAL DATA VIEWER: ON THE COLORADO STATE UNIVERSITY WEBSITE HTTP://BIOCYCLE.ATMOS.COLOSTATE.EDU/SHINY/MILANKOVITCH/). THE MINI-APP IS VERY GOOD IN THAT IT ALLOWS THE CALCULATION OF ORBITAL DATA WITH VARIATION IN TIME INPUT. THIS ORBITAL DATA IS THEN USED TO CALCULATE TEMPERATURE RELATED MEASUREMENTS.*

All the climate changes caused by earth's orbital changes are called Milankovitch Cycles named after Serbian geophysicist and astronomer Milutin Milanković (May 28, 1879 - December 12, 1958).

***It is critical to note that these cyclical changes do not lead to <u>net</u> climate change.***

# ATMOSPHERIC CHANGES

Volcanic events can inject large amounts of particulate and gasses into the stratosphere and cause near global-scale cooling. These effects are generally limited by the length of volcanic activity and the particulate and aerosol settling time (in the order of several years). Mount Pinatubo in the Philippines, erupted on June 15, 1991 and its impact was extensively studied by scientists around the world[26]. The vivid red sunsets around the globe were a result primarily of sulfate aerosol from the reaction of $SO_2$ gas released in the eruption.

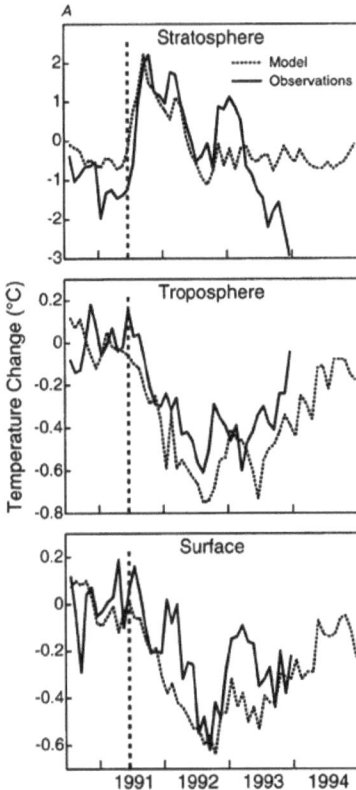

"Effects on climate were an observed surface cooling in the Northern Hemisphere of up to 0.5 to 0.6°C, equivalent to a hemispheric-wide reduction in net radiation of 4 watts per square meter and a cooling of perhaps as large as -0.4°C over large parts of the Earth in 1992-93."[27] The temperature profiles to the left are from Figure 12A[28].

*MEASURED AND MODELLED TEMPERATURES FOR ATMOSPHERIC LAYERS AFTER MT. PINATUBO ERUPTION (SOURCE: FIGURE 12A IN SELF ET AL, FOOTNOTE 24)*

[26] **Fire and Mud: Eruptions and Lahars of Mount Pinatubo, Philippines.** Edited by Christopher G. Newhall and Raymundo S. Punongbayan. University of Washington Press, Seattle and London, 1996.

[27] Stephen Self, Jing-Xia Zhao, Rick E. Holasek, Ronnie C. Torres, and Alan J. King. "The Atmospheric Impact of the 1991 Mount Pinatubo Eruption" in **Fire and Mud** (See footnote 22).

[28] Self *et al.* "Observed and modeled monthly temperature change of stratosphere, troposphere, and surface after the Mount Pinatubo eruption. Stratospheric observations are 30-mbar zonal mean temperature at 10° S. lat; model results are 10- to 70-mbar layer at 8° to 16° S. lat. Other results are essentially global, with observed surface temperature derived from a network of meteorological stations. Base period for tropospheric temperatures is 1978-92, while troposphere and surface are referenced to the 12 months preceding the Pinatubo eruption, the latter marked by a vertical dashed line."

Ocean heating or cooling events such as El Niño and La Niña (complex weather patterns resulting from variations in ocean temperatures in the Equatorial Pacific) have about the same magnitude of global atmospheric temperature change as large individual volcanic events and last a similar short time until there is a global rebound onto the main curve. There is most certainly a coupling of global atmospheric temperature to ocean temperature and the oceans are indeed a heat sink but the coupling is complex and the mixing of ocean water from the surface to deep ocean requires significant time.

What is important to remember and to focus upon is the trend in the overall temperature curve over the very recent past.

Volcanos and ocean currents (mentioned above) are not part of the particular problems of current interest in global climate change although they do affect global climate but usually only for a relatively short time. **However, as the deep ocean becomes warmer any attempts at halting or reversing global atmospheric warming will become moot**.

There is an important fundamental notion to emphasize with respect to ocean temperature changes and their effect on the atmospheric temperature. From the outset, any calculation (or, in the case of global warming nay-sayers, wild speculation) must take into account the source of the ocean heating not just locally but globally. The ocean does not heat itself! Therefore, the basic issue will always end up being the total radiation flux to and from the planet through the atmosphere.

GREENHOUSE GASES

It is the greenhouse gases[29] and, most particularly, carbon dioxide ($CO_2$), that are the significant problem in global warming.

Svante August Arrhenius (Swedish, 1859-1927) received the Nobel Prize in Chemistry in 1903 for his work in electrochemistry. In 1896, he was the first to suggest that fossil fuel combustion may eventually result in enhanced global warming. He formulated the relationship between atmospheric carbon dioxide concentrations and temperature[30]. This was based upon the infrared absorption of water vapor and carbon dioxide. Arrhenius suggested a doubling of the $CO_2$ concentration would lead to a 5°C temperature rise.

These early projections of fossil fuel effects on climate were not taken up by the general scientific community or there was a wide-spread belief that man-made material could not be a significant contributor to atmospheric change. Water vapor was seen as a much more influential greenhouse gas.

---

[29] The greenhouse gases are water vapor, carbon dioxide ($CO_2$), methane ($CH_4$), nitrous oxide ($N_2O$) and ozone ($O_3$).
[30] http://www.lenntech.com/greenhouse-effect/global-warming-history.htm#ixzz3cls4dGpB. Retrieved 06/11/2015.

This viewpoint is understandable since water vapor is the essential component of the thermodynamic "heat engine"[31] that powers the motions of the atmosphere (basically - weather).  Certainly, water vapor is a greenhouse gas but it has a self-limiting value that is dictated by overall temperature and atmospheric capacity – that is, it can't get higher than 100% RH (relative humidity).  In terms of basic thermodynamics and atmospheric modeling this is a bit simplistic but not considering water vapor in this discussion is justified scientifically and by the fact that it is anthropogenic change that we are interested in examining.

The effects of increasing $CO_2$ concentrations on global warming are straightforward and thermodynamically undeniable.  The simple curve of increasing atmospheric $CO_2$ shows a very clean and reasonably stable increase in the last 30 years.

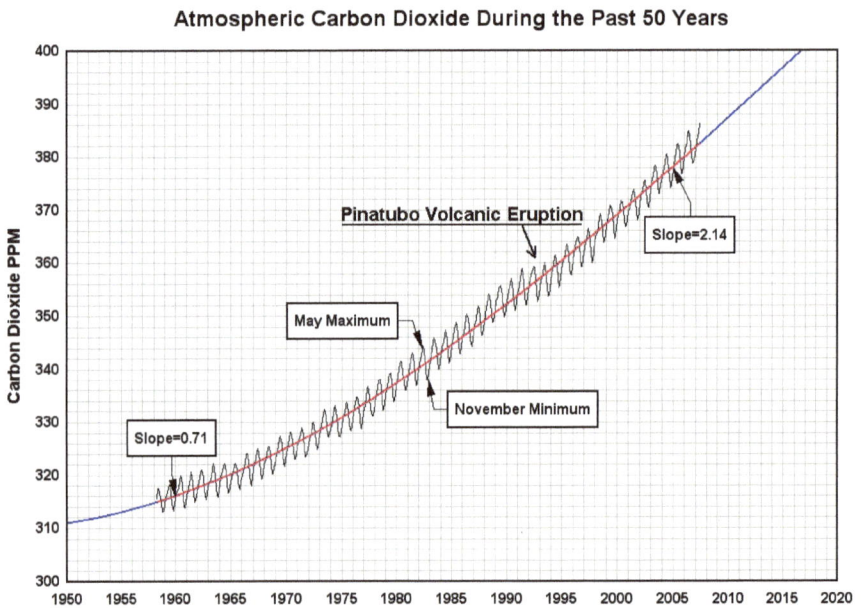

### Atmospheric Carbon Dioxide During the Past 50 Years

*50 YEARS OF $CO_2$ MEASUREMENTS (ANNOTATIONS FROM HTTP://WWW.PLANETFORLIFE.COM/CO2HISTORY/ RETRIEVED 06/12/2015. THE DATA IS NOAA DATA UNALTERED AVAILABLE AT HTTP://WWW.ESRL.NOAA.GOV/GMD/CCGG/TRENDS/.  THE DATA IS RECORDED AT THE MAUNA LOA OBSERVATORY, HAWAII AND THAT INSURES THAT IT IS NOT COMPLICATED BY VARIABLE LOCAL SOURCES.)*

---

[31] Laliberté F., Zika J., Mudryk L., Kushner P.J., Kjellsson J., Döös K.  "Atmospheric dynamics.  Constrained work output of the moist atmospheric heat engine in a warming climate." **Science.** 2015 Jan 30; 347(6221): 540-3.

Data is available for atmospheric $CO_2$ prior to modern direct atmospheric measurements that were initiated in 1958 by Charles David Keeling, of Scripps Institution of Oceanography at the University of California San Diego. He made frequent regular measurements at the South Pole and in Hawaii. In a paper from 1960 Keeling and associates already had enough data to conclude that the rise in atmospheric $CO_2$ was roughly equivalent to increases in fossil fuel use[32]. The records of atmospheric $CO_2$ for millennia prior to 1958 are derived from extensive measurements of trapped gas bubbles in Arctic and Antarctic ice cores. This data is presented below.

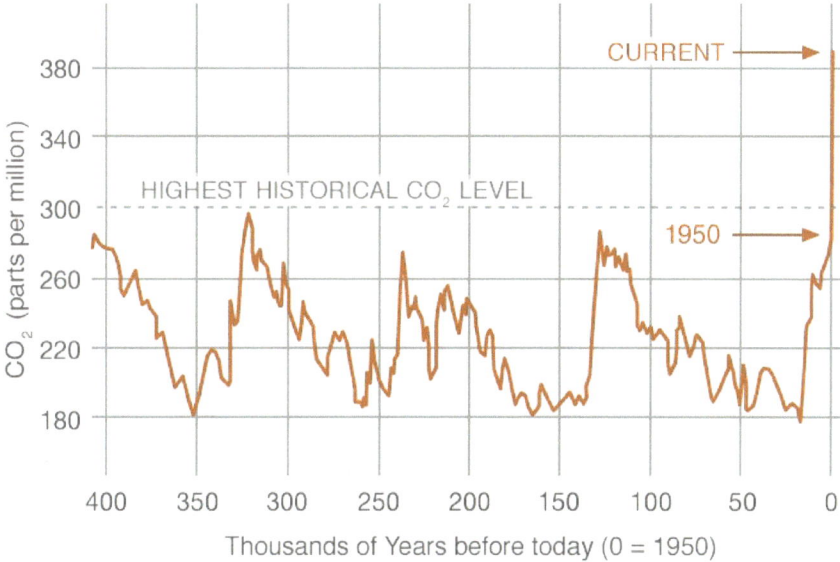

THIS DIAGRAM WAS TITLED "PROXY (INDIRECT) MEASUREMENTS" AND HAD THE LEGEND INDICATING THAT THE VALUES WERE RECONSTRUCTED FROM ICE CORES – SOURCE: NOAA. RETRIEVED 06/12/2015 FROM HTTP://CLIMATE.NASA.GOV/VITAL-SIGNS/CARBON-DIOXIDE/. THE DATA POINT AFTER 1950 MARKED "CURRENT" REFERS TO LAST 2010 VALUE FROM THE DIRECT MEASUREMENTS (PRIOR DIAGRAM). THIS SHOWS 3 ICE AGES [THESE ARE REPRESENTED BY THE LOW VALUES OF $CO_2$ - THE 180 PPM LINE].

---

[32] C. D. Keeling, "The Concentration and Isotopic Abundances of Carbon Dioxide in the Atmosphere", **Tellus**. 12, 200-203, 1960.

## RECENT MONTHLY MEAN CO$_2$ AT MAUNA LOA

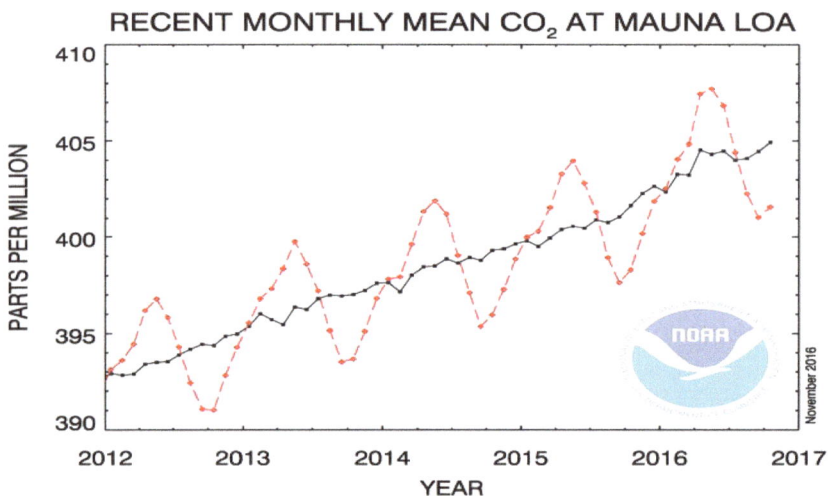

http://www.esrl.noaa.gov/gmd/ccgg/trends/index.html

The CO$_2$ measurement for September 2016 was **404.42 ppm** (retrieved 11/10/2016 from http://climate.nasa.gov/vital-signs/carbon-dioxide/)[33].

This datum is absolutely on the same curve with the same slope as presented in the 50-year plot above. This should be alarming given the consensus held among scientists on global warming. *Unfortunately, scientists do not control the world, businessmen and politicians do! Politicians rely on voters, the "ordinary people" who can make the difference if they are properly informed and motivated.*

The essential point to grasp is that the greenhouse gases (GHGs) trap outgoing radiation from the planet that would normally lead to planetary cooling. Although there are many other GHGs, methane being one of the most significant, it is critical to focus on the primary culprit, CO$_2$ hence – fossil fuel burning.

---

[33] The home page of the NASA website is http://climate.nasa.gov/. It is an excellent and accessible presentation of data from reliable sources and also indicates the overwhelming scientific consensus that global warming is a direct result of human activity. The American Chemical Society (ACS) also has an accessible and instructive Climate Science Toolkit (see http://www.acs.org/content/acs/en/climatescience.html) that explains everything from radiation and absorption of heat to greenhouse gases.

It is stunning how many nonsense pages of anti-global warming rhetoric and misleading data have been posted on the Internet. As a scientist, I am absolutely in favor of spirited discussion and debate over new findings and data but it is insulting to believe that people with virtually no qualifications and apparently less vision can cause so much confusion. The Flat Earth Society[34] exists but no-one credible turns to them for navigational tips!

Besides those who obfuscate why would it be difficult to accept the correlation of anthropogenic $CO_2$ (fossil fuel burning), atmospheric $CO_2$ increases and global warming? One suggestion is that it is the age-old problem of statistics versus the perception of the individual event. It is very difficult to accept global warming projections if you live in Winnipeg, Manitoba, Canada and get polar storms and sub-zero temperatures in June. Another, and more likely possibility, is that there are vested national economic interests in trying to point the blame away from fossil fuel burning. None-the-less, the data and the trending are abundantly clear. Although correlation does not prove causality there is an abundance of subsidiary evidence and good science that makes a virtually irrefutable case for $CO_2$ as global enemy #1. Just remember the "scientists" working for the cigarette companies denying that smoking was causing increases in the rate of lung cancer.

THE CARBON CYCLE

Besides the statistical problem of perception indicated above there is a real overlay of complexity that does exist. The carbon cycle is an essential "life cycle" on the planet. Carbon is cycled, from its inorganic origins in the atmosphere as $CO_2$, by the oceans into carbonates that form reefs, limestone, marble and other composites and by plants into cellulose, glucose and many other organic molecules that are life itself. In fact, as has already been discussed, it is the recycling of organic matter by burning that is the particular problem of global warming.

The available literature on the carbon cycle is vast and cannot be covered here but there are some basic points that should be considered:

- the earth has a "buffering capacity" and threshold limits for coming to an equilibrium with many natural substances
- there are many "sinks" for $CO_2$ but the rate of conversion is an issue - if the input levels are too high there will be problems. This is actually an expansion of the first and fundamental consideration.

Although there are vast potential sinks for $CO_2$ the net production rate is clearly exceeding the rate of conversion.

---

[34] http://www.theflatearthsociety.org/cms/. Retrieved 06/13/2015.

## $CO_2$ SINKS AND DEEPER CONSIDERATIONS

The ocean itself is a giant sink for atmospheric $CO_2$ but there are severe consequences in the acidification of the oceans. This relationship is very well captured in the figure below.

Dissolved **$CO_2$** and Ocean Acidity (**pH**)

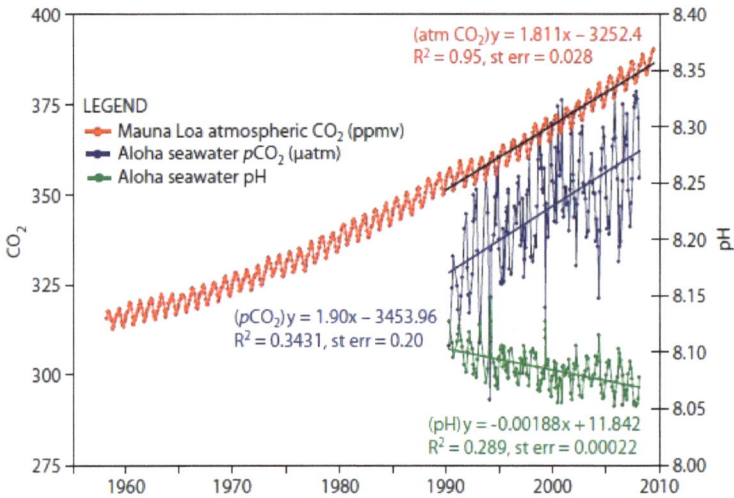

LEGEND
- Mauna Loa atmospheric $CO_2$ (ppmv)
- Aloha seawater $pCO_2$ (µatm)
- Aloha seawater pH

(atm $CO_2$)y = 1.811x − 3252.4
$R^2$ = 0.95, st err = 0.028

($pCO_2$)y = 1.90x − 3453.96
$R^2$ = 0.3431, st err = 0.20

(pH)y = −0.00188x + 11.842
$R^2$ = 0.289, st err = 0.00022

*"THIS FIGURE SHOWS THE CORRELATION BETWEEN RISING LEVELS OF CARBON DIOXIDE ($CO_2$) IN THE ATMOSPHERE AT MAUNA LOA (RED) WITH RISING LEVELS OF $CO_2$ DISSOLVED IN THE OCEAN AT NEARBY STATION ALOHA (BLUE), AND THE CONSEQUENT INCREASE IN ACIDITY OF THE OCEAN THAT IS SEEN AS A DECREASE IN OCEAN PH (GREEN, RIGHT SCALE). GLOBAL WARMING DEALS A DOUBLE BLOW TO THE OCEANS BECAUSE NOT ONLY DOES THE OCEAN WARM BY ABSORBING PART OF THE ATMOSPHERIC TEMPERATURE INCREASE, IT ALSO ABSORBS ABOUT ONE-THIRD OF THE INCREASE IN $CO_2$, WHICH REACTS WITH WATER TO PRODUCE CARBONIC ACID." THE ENTIRE FIGURE AND CAPTION ARE A VERY GOOD SUMMARY OF THE SOURCE / SINK / CONSEQUENCE CORRELATION. IT WAS RETRIEVED ON 06/13/2015 – ORIGINAL FIGURE A2 ON THE WEBSITE HTTP://WWW.SKEPTICALSCIENCE.COM/GLOBAL-WARMING-IN-A-NUTSHELL.HTML.*

Now it is appropriate to map the irrefutable increase in atmospheric $CO_2$ to the actual increase in average global temperature but keeping in mind that the ocean has been slowing down the warming by capturing atmospheric heat and $CO_2$ at the potentially grave cost of ocean acidification.

28

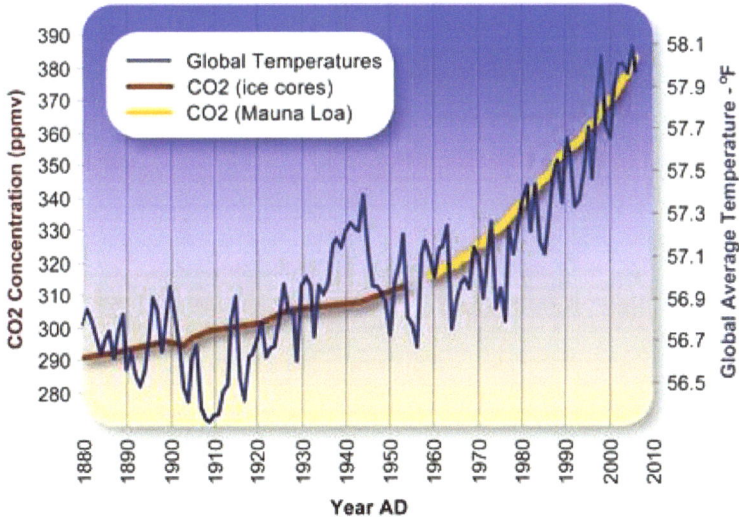

## Global Average Temperature and Carbon Dioxide Concentrations, 1880 - 2006

Legend:
- Global Temperatures
- CO2 (ice cores)
- CO2 (Mauna Loa)

CO2 Concentration (ppmv): 280, 290, 300, 310, 320, 330, 340, 350, 360, 370, 380, 390

Global Average Temperature - °F: 56.5, 56.7, 56.9, 57.1, 57.3, 57.5, 57.7, 57.9, 58.1

Year AD: 1880, 1890, 1900, 1910, 1920, 1930, 1940, 1950, 1960, 1970, 1980, 1990, 2000, 2010

Data Source Temperature: ftp://ftp.ncdc.noaa.gov/pub/data/anomalies/annual.land_and_ocean.90S.90N.df_1901-2000mean.dat
Data Source CO2 (Siple Ice Cores): http://cdiac.esd.ornl.gov/ftp/trends/co2/siple2.013
Data Source CO2 (Mauna Loa): http://cdiac.esd.ornl.gov/ftp/trends/co2/maunaloa.co2
& http://www.esrl.noaa.gov/gmd/webdata/ccgg/trends/co2_mm_mlo.dat

Graphic Design: Michael Ernst, The Woods Hole Research Center

*THE PLOT ABOVE WAS RETRIEVED 06/13/2015 FROM HTTP://WWW.WHRC.ORG/RESOURCES/PRIMER_FUNDAMENTALS.HTML.*

IRREFUTABLE FACTS

1. Global warming is real.
2. Global warming is highly correlated with increasing atmosphere $CO_2$ levels.
3. Increasing atmospheric $CO_2$ levels are highly correlated to anthropogenic fossil fuel combustion.
4. Any residual climate change skeptics are either misinformed or uninformed.

We should consider the onset of the global warming "crisis" as the industrial revolution. The fundamental change from agrarian (agriculturally based) society to industrial consumption and production. If one follows the economic development of nations from as early as 1760[35] through to today it is virtually impossible to decouple energy use and production from economic parameters. Some of these issues will be covered in the next chapter.

---

[35] It is difficult to fix any specific date to the beginning of the industrial revolution because it cannot really be considered a revolution but a gradual evolution that picked up pace in Great Britain in the mid-nineteenth century. There is a considerable volume of historical and anthropological literature that deals with the transition from agrarian economies to those based on trade and industry.

# WORLD ENERGY CONSUMPTION & PRODUCTION

*"The sooner we get started with alternative energy sources and recognize that fossil fuels make us less secure as a nation, and more dangerous as a planet, the better off we'll be."*

Lindsey Olin Graham is an American politician and member of the Republican Party, who has served as the senior United States Senator from South Carolina since 2003. On June 1, 2015, he announced his candidacy for the Presidency in the 2016 election.

## CONSUMPTION OF ENERGY

Although it would be academically admirable to look at energy consumption for every country in the world it would be a real waste of time and energy[36]. Looking at the top producers will give us a nominally (scientifically / statistically) complete answer to help shape solutions for climate change.

**Primary Energy Consumption (Quadrillion BTU)**
**Rankings based on 2014**

|  | 2010 | 2011 | 2012 | 2013 | 2014 | 2014% | 2014% |
|---|---|---|---|---|---|---|---|
| **WORLD** | **516.18** | **529.81** | **539.57** | **537.28** | **536.65** | **100.00%** | Cumulative |
| China | 102.79 | 112.39 | 118.28 | 118.64 | 115.03 | 21.44% | 21.44% |
| United States | 97.44 | 96.84 | 94.41 | 97.14 | 98.30 | 18.32% | 39.75% |
| Russia | 29.86 | 30.49 | 31.22 | 30.33 | 30.68 | 5.72% | 45.47% |
| India | 23.58 | 24.73 | 26.06 | 23.51 | 24.31 | 4.53% | 50.00% |
| Japan | 21.62 | 20.94 | 20.49 | 19.56 | 19.10 | 3.56% | 53.56% |
| Canada | 13.63 | 14.09 | 14.04 | 14.41 | 14.55 | 2.71% | 56.27% |
| Germany | 14.02 | 13.46 | 13.57 | 13.49 | 13.07 | 2.44% | 58.70% |
| Brazil | 11.42 | 11.84 | 12.20 | 12.64 | 12.80 | 2.39% | 61.09% |
| Korea, South | 10.98 | 11.41 | 11.60 | 11.01 | 11.11 | 2.07% | 63.16% |
| Iran | 9.20 | 9.48 | 9.76 | 10.16 | 10.71 | 2.00% | 65.16% |
| Saudi Arabia | 8.40 | 8.86 | 9.66 | 9.86 | 10.32 | 1.92% | 67.08% |
| France | 10.97 | 10.76 | 10.61 | 10.62 | 10.21 | 1.90% | 68.98% |
| United Kingdom | 8.95 | 8.48 | 8.70 | 8.48 | 8.02 | 1.49% | 70.47% |
| Mexico | 7.25 | 7.51 | 7.39 | 7.49 | 7.50 | 1.40% | 71.87% |
| Indonesia | 6.31 | 6.41 | 6.99 | 6.96 | 7.30 | 1.36% | 73.23% |
| **TOP 15 in 2014** |  |  |  |  |  | 73.23% |  |

Primary Source:  U.S. Energy Information Administration – retrieved 11/12/2016 from
http://www.eia.gov/beta/international/data/browser/#/?pa=000000001&c=ruvvvvvfvtvnvv1urvvvvfvvvvvvfvvvou
20evvvvvvvvvvnvvuvo&ct=0&vs=INTL.44-2-AFG-BTU.A&ord=SA&cy=2014&vo=0&v=H&start=2010&end=2014.
Last column - Percentages added by author.

---

[36] The pun is intended … It is a very short list of countries that produce and consume a major portion of energy. Also, it borders on the near impossible to get accurate numbers from many countries lacking either infrastructure or political will to exhibit transparency.

The top ranked 5 countries in the table account for 53.56% of global energy consumption (2014 data). When looking at consumption of energy it is important to scale that consumption in a number of ways to better understand various aspects of the problem. The table above presents energy consumption by country and indicates % of the world consumption in the last column. Another way to look at the data is the *per capita* consumption and that provides a very different view of the use patterns and may provide interesting input for general discussions on energy production and use (see the Table below).

### Per Capita Energy Use

| Country | 2009 | 2010 | 2011 | 2012 | 2013 | World Rank for 2013 |
|---|---|---|---|---|---|---|
| | Energy use (kg of oil equivalent per capita) | | | | | |
| Qatar | 15230.37 | 15657.01 | 16424.58 | 18899.36 | 19120.34 | 1 |
| Iceland | 16911.08 | 17023.17 | 18157.60 | 17630.07 | 18177.25 | 2 |
| Trinidad and Tobago | 14203.09 | 15109.24 | 14791.34 | 14346.94 | 14537.57 | 3 |
| Curacao | 15722.02 | 13815.73 | 16162.03 | 13441.62 | 11800.98 | 4 |
| Bahrain | 10212.89 | 9952.24 | 9628.85 | 9417.04 | 10171.68 | 5 |
| Kuwait | 10947.55 | 10525.89 | 10040.44 | 10121.44 | 9757.45 | 6 |
| United Arab Emirates | 7780.36 | 7414.66 | 7361.15 | 7552.52 | 7691.01 | 7 |
| Brunei Darussalam | 7860.92 | 8238.79 | 9695.71 | 9444.96 | 7392.87 | 8 |
| Luxembourg | 7951.72 | 8329.43 | 8056.36 | 7721.96 | 7310.31 | 9 |
| Canada | 7432.23 | 7391.73 | 7484.16 | 7259.24 | 7202.23 | 10 |
| United States | 7056.78 | 7161.52 | 7029.25 | 6812.49 | 6915.84 | 11 |
| Russian Federation | 4531.29 | 4827.81 | 5057.52 | 5174.40 | 5093.06 | 20 |
| Japan | 3688.62 | 3895.68 | 3614.38 | 3543.27 | 3570.44 | 32 |
| China | 1692.68 | 1845.74 | 1994.40 | 2079.12 | 2226.27 | 53 |
| India | 545.26 | 562.70 | 574.32 | 595.10 | 606.05 | 112 |

*World Bank Last update 02/01/2017*
*http://data.worldbank.org/indicator/EG.USE.PCAP.KG.OE*
*Rankings provided by author*
*The last complete data set represented in the table is for 2013.*

The United States comes in at #11 and the Russian Federation is not far behind at #20. Japan (#32) has substantially lower *per capita* use than the United States – only about 51.6% of the US rate. China is another significant step down at #53 and 32.2% of the US. India has a world rank of 112 and is at 8.8% of the US.

Seeing Qatar, the United Arab Emirates, Kuwait, and Bahrain at the top of the list is not surprising in the least since they are major oil producing nations where consumption is cheap and the relative incomes are high. Trinidad and Tobago produces a relatively large quantity of petroleum.

The *per capita* consumption table has two seemingly strange countries near the top of the list – Curacao and Luxembourg. Curacao produces no significant power from reserves of its own but its economy is driven by oil refining since as early as 1918.[37] Luxembourg is also not a net energy producer but has a very diversified economy:

> "The economy of Luxembourg is largely dependent on the banking, steel, and industrial sectors. Luxembourgers enjoy the second highest *per capita* gross domestic product in the world (CIA 2007 est.), behind Qatar. Luxembourg is seen as a diversified industrialized nation, contrasting the oil boom in Qatar, the major monetary source of the southwest Asian state."
>
> https://en.wikipedia.org/wiki/Economy_of_Luxembourg

Iceland is high up in the list and is there primarily because of its enviable situation of having access to significant geothermal resources. About 85% of total primary energy supply in Iceland is derived from domestically produced renewable energy sources[38]. Another factor to keep in mind is that Iceland has a small population and total energy consumption in 2102 was ranked at 100th or about 0.04% of the world total. Nice job but not a significant player!

Just by briefly considering the two prior tables and the tentative explanations for *per capita* rankings when proposing solutions for climate change it is indeed important to consider:

- Energy consumption *per capita*
- Emissions ($CO_2$)
- Energy production (partitioned by source).

---

[37] "The discovery of oil in the Maracaibo Basin of Venezuela in the early 20th century forced the Venezuelan government to search for ideal locations for large scale refining. Curaçao's proximity to the country, naturally deep harbors, and stable government led Royal Dutch Shell to construct the Isla Refinery, the largest refinery in the world at the time. Presently Venezuela's state oil company, Petróleos de Venezuela (PDVSA) operates the Isla refinery, which has a 320,000 barrel per day capacity." Retrieved 2025/2017. https://en.wikipedia.org/wiki/Economy_of_Cura%C3%A7ao.

[38] The Independent Icelandic Energy Portal – http://askjaenergy.org/iceland-introduction/iceland-energy-sector/. Retrieved 06/16/2015.

Before moving on to production a very visually clear display of consumption is presented in a satellite view of the world at night[39].

Certainly, data for energy consumption is absolutely comprehensible in terms of region but not in terms of population density. The primary variable is economics...wealth.

[39] Retrieved 06/16/2015 from http://geology.com/articles/satellite-photo-earth-at-night.shtml. The image was compiled using data from a sensor aboard the NASA-NOAA Suomi National Polar-orbiting Partnership satellite launched in 2011. More area-specific and higher resolution images are available.

## CO$_2$ EMISSIONS

Energy consumption, in most countries in the world, is intimately linked with fossil fuel burning hence, production of CO$_2$.

## CO$_2$ Emissions from Energy Consumption (Million Metric Tons)
### Rankings based on 2014

| | 2010 | 2011 | 2012 | 2013 | 2014 | 2014% | 2014% |
|---|---|---|---|---|---|---|---|
| WORLD | 31671.05 | 32750.23 | 25896.39 | 32689.82 | 32286.31 | 100.00% | Cumulative |
| China | 7895 | 8703 | 8964 | 8891 | 8340 | 25.83% | 25.83% |
| United States | 5578.3 | 5479.71 | 5270 | 5364 | 5412 | 16.76% | 42.59% |
| India | 1761.8 | 1798.7 | | 1705.5 | 1772.5 | 5.49% | 48.08% |
| Russia | 1669 | 1707.95 | | 1705 | 1737 | 5.38% | 53.46% |
| Japan | 1155 | 1194 | 1252 | 1184 | 1158 | 3.59% | 57.05% |
| Germany | 795.6 | 780.75 | 786.6 | 774.5 | 742.5 | 2.30% | 59.35% |
| Iran | 564.3 | 591.58 | | 618.6 | 645.7 | 2.00% | 61.35% |
| Korea, South | 599 | 661 | 667 | 631 | 631 | 1.95% | 63.30% |
| Canada | 585.8 | 597.01 | 591.3 | 599.4 | 604.7 | 1.87% | 65.18% |
| Saudi Arabia | 511 | 551 | 584 | 550 | 576.02 | 1.78% | 66.96% |
| Brazil | 458.4 | 474.23 | | 525.4 | 544 | 1.68% | 68.65% |
| Indonesia | 426.9 | 454.57 | | 506 | 537 | 1.66% | 70.31% |
| South Africa | 463.9 | 452.8 | 471.2 | 454.6 | 454.39 | 1.41% | 71.72% |
| United Kingdom | 532.4 | 492.15 | 507.9 | 480 | 442 | 1.37% | 73.09% |
| Mexico | 435 | 432.73 | 430.8 | 435.8 | 433.8 | 1.34% | 74.43% |
| **Top 15 in 2014** | | | | | | 74.43% | |

Primary Source: U.S. Energy Information Administration – retrieved 11/10/2016 from

http://www.eia.gov/beta/international/data/browser/#/?pa=00000000000000000000000002&c=ruvvv vvfvtvnvv1urvvvvfvvvvvvfvvvou20evvvvvvvvvvvnvvuvo&ct=0&vs=INTL.44-8-AFG-MMTCD.A&ord=SA&vo=0&v=H&start=2007&end=2014. Percentages added by author.

The top ranked 5 countries in the table account for 57.05% of total CO$_2$ emissions from energy consumption (2014 data). This should be no surprise to anyone except the most die-hard climate change skeptics. The largest consumers of energy are the largest producers of CO$_2$ – HENCE THE LARGEST CONTRIBUTORS TO GLOBAL WARMING. One does not have to be a conspiracy theorist to propose that the biggest users are the biggest abusers and are the least likely to make substantial changes in the *status quo* unless forced. Part of the query in this book is what would be sufficient "force" or, perhaps a better term would be, "inducement" to change their energy consumption profiles in significant and meaningful ways.

# PRODUCTION OF ENERGY

## Total Primary Energy Production
## (Quadrillion BTU)

| | 2010 | 2011 | 2012 | 2013 | 2014 | 2014% | 2014% |
|---|---|---|---|---|---|---|---|
| WORLD | 509.46 | 524.86 | 535.98 | 536.74 | 545.59 | 100.00% | Cumulative |
| China | 88.28 | 94.57 | 98.89 | 103.90 | 103.99 | 19.06% | 19.06% |
| United States | 74.73 | 77.91 | 79.10 | 81.68 | 87.57 | 16.05% | 35.11% |
| Russia | 52.83 | 54.33 | 55.07 | 55.40 | 55.33 | 10.14% | 45.25% |
| Saudi Arabia | 25.43 | 26.72 | 27.94 | 27.60 | 27.59 | 5.06% | 50.31% |
| Canada | 18.33 | 18.74 | 19.13 | 19.71 | 20.54 | 3.76% | 54.07% |
| Indonesia | 13.84 | 15.48 | 16.35 | 15.42 | 14.95 | 2.74% | 56.81% |
| India | 15.72 | 16.09 | 16.10 | 14.73 | 15.52 | 2.84% | 59.66% |
| Australia | 12.93 | 12.91 | 13.34 | 13.36 | 14.22 | 2.61% | 62.26% |
| Iran | 14.61 | 14.79 | 13.63 | 13.04 | 13.89 | 2.55% | 64.81% |
| Qatar | 8.05 | 9.53 | 9.91 | 10.21 | 10.01 | 1.84% | 66.64% |
| Brazil | 9.46 | 9.90 | 9.76 | 9.71 | 10.08 | 1.85% | 68.49% |
| United Arab Emirates | 7.64 | 8.46 | 8.92 | 8.99 | 9.02 | 1.65% | 70.15% |
| Norway | 9.46 | 9.07 | 9.47 | 8.96 | 9.12 | 1.67% | 71.82% |
| Mexico | 8.75 | 8.81 | 8.63 | 8.55 | 8.32 | 1.53% | 73.34% |
| Venezuela | 7.27 | 7.81 | 7.58 | 7.51 | 7.60 | 1.39% | 74.73% |
| TOP 15 in 2014 | | | | | | 74.73% | |

Primary Source:  U.S. Energy Information Administration – retrieved 11/12/2016 from
http://www.eia.gov/beta/international/data/browser/#/?pa=004&c=ruvvvvvfvtvnvv1urvvvvfvvvvvvfvvvou20evvvv
vvvvvnvvuvo&ct=0&vs=INTL.44-1-AFG-QBTU.A&ord=SA&cy=2010&vo=0&v=H&start=2009&end=2014.
Percentages added by author.

| Country | 2012 Rankings | | |
| --- | --- | --- | --- |
| | Energy Consumption | $CO_2$ Emissions | Energy Production |
| China | 1 | 1 | 1 |
| United States | 2 | 2 | 2 |
| Russia | 3 | 4 | 3 |
| India | 4 | 3 | 7 |
| Japan | 5 | 5 | 42 |
| Germany | 6 | 6 | 25 |
| Canada | 7 | 10 | 5 |
| Brazil | 8 | 11 | 11 |
| Korea, South | 9 | 7 | 43 |
| France | 10 | 18 | 23 |
| Iran | 11 | 8 | 8 |
| Saudi Arabia | 12 | 9 | 4 |
| United Kingdom | 13 | 12 | 24 |
| Mexico | 14 | 15 | 13 |
| Italy | 15 | 17 | 46 |
| South Africa | 19 | 13 | 20 |
| Idonesia | 16 | 14 | 6 |
| Australia | 18 | 16 | 9 |
| Qatar | 40 | 40 | 10 |
| United Arab Emirates | 26 | 26 | 14 |
| Venezuela | 29 | 32 | 15 |
| | Large scale importer | | |
| Ranked in top 15 other than in Consumption | | | |

*Percentages added by author. Primary Source: U.S. Energy Information Administration. The author has generated rankings based on the raw data for 2012 and the table annotations.*

Some conclusions follow easily and logically from looking at the rankings in the table above. By looking at Consumption, $CO_2$ Emission, and Production we can get some pretty good ideas about the energy sources and distributions. For the first 3 countries – China, United States and Russia there is little surprise. They are all major producers and consumers of fossil fuels. India produces a lot of energy and consumes a great deal and has emissions slightly larger than would be expected from its consumption leading to a possible analysis that pollution controls and standards are lax and that much consumption is not at an industrial level. Japan is clearly a net importer of fossil fuels and has emissions congruent with consumption. That is the simplest and most accurate assessment of the global issues under consideration – the first 5 countries in the tables represent 52.79% of Energy Consumption, 56.48% of $CO_2$ Emissions and 52.69% of Energy Production. There is no coincidence in the correspondence of these values nor should there be any hesitation in finger pointing and blame.

**Five countries (China, United States, India, Russia, Japan) are responsible for more than half of the global climate crisis.**

It is also far too tempting at this point to indicate that three of them, the top three producers, consumers, polluters and culprits are 3 of 5 United Nations Security Council permanent members. This should be real material for conspiracy theorists although, for the life of me, I could not come up with any conspiracy other than "a confederacy of dunces"[40].

**Top ten annual net oil importers, 2014**
million barrels per day

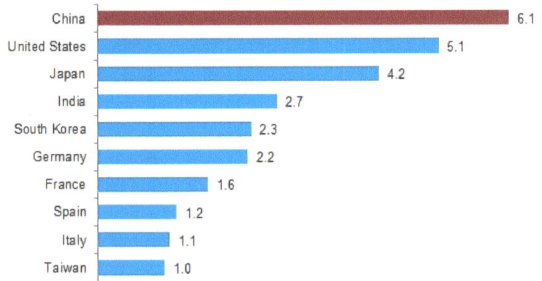

Despite high levels of production of energy, all top consumers (except Japan), are large-scale importers of oil (except Russia) as seen in the histogram.

HTTP://WWW.EIA.GOV/BETA/INTERNATIONAL/ ANALYSIS.CFM?ISO=CHN. RETRIEVED 06/17/2015 (SAME AS 11/17/2016)

| | |
|---|---|
| China | 6.1 |
| United States | 5.1 |
| Japan | 4.2 |
| India | 2.7 |
| South Korea | 2.3 |
| Germany | 2.2 |
| France | 1.6 |
| Spain | 1.2 |
| Italy | 1.1 |
| Taiwan | 1.0 |

eia  Note: Estimates of total production less consumption. Does not account for stock build.
Source: U.S. Energy Information Administration, *Short-Term Energy Outlook*, May 2015

**China's natural gas production and consumption, 2000-2013**
trillion cubic feet

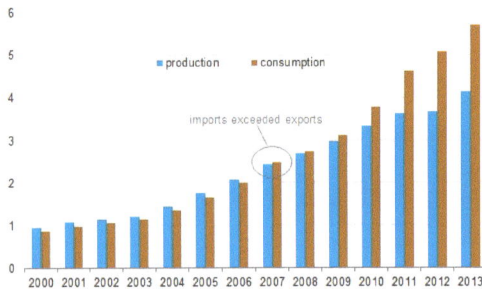

■ production  ■ consumption

imports exceeded exports

eia  Source: U.S. Energy Information Administration, *International Energy Statistics*

In addition to being the top oil importer China has gone from a net producer of natural gas to an importer (2007) and the size of imports seems to be steadily growing.

HTTP://WWW.EIA.GOV/BETA/INTERNATIONAL/AN ALYSIS.CFM?ISO=CHN RETRIEVED 06/17/2015.

---

[40] *A Confederacy of Dunces* is an episodic novel written by John Kennedy Toole which appeared in 1980, eleven years after Toole's suicide. Toole earned a posthumous Pulitzer Prize for Fiction in 1981.

This is rapidly brought into context by considering the distribution of fuel sources used by China in electricity generation as shown in the pie chart below.

The previous graphs of oil and natural gas are really insignificant in this diagram since they represent only about 2% and 4% respectively of the fuels for generation. Coal is the primary power source in China. Comparing Russia is quite different and although Russia has very large coal reserves the primary fuel is natural gas – albeit, a fossil fuel, but intrinsically the cleanest of all of them. Natural gas is generally much easier to access than coal and transportation and handling can be performed more efficiently and cost-effectively. This is the logical and likely reason for the predominance of natural gas in Russia's consumption pattern.

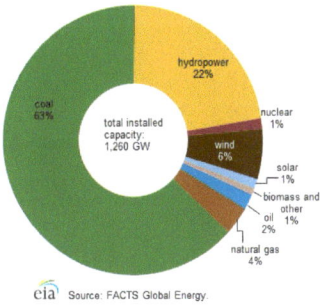

China's installed electricity capacity share by fuel, end 2013

eia  Source: FACTS Global Energy.

HTTP://WWW.EIA.GOV/BETA/INTERNATIONAL/AN ALYSIS.CFM?ISO=CHN RETRIEVED 06/17/2015.

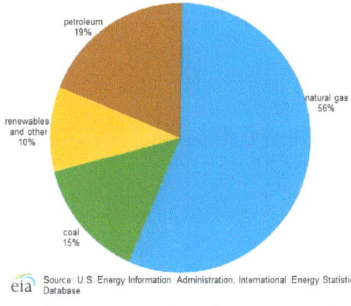

Russia's primary energy consumption, 2011

eia  Source: U.S. Energy Information Administration, International Energy Statistics Database

HTTP://WWW.EIA.GOV/BETA/INTERNATIONAL/ANAL YSIS.CFM?ISO=RUS RETRIEVED 06/18/2015.

For many in North America and Europe it may come as a surprise but coal, as it had been in 19th century England, is still a very large contributor to $CO_2$ emissions and global warming, thanks primarily to the fact that China has become a major producer of energy and much of that is from combustion of coal.

Anyone who did atmospheric research[41] on Beijing in preparation for the Summer Olympics of 2008 should have been seriously concerned about extremely high particulate air pollution events in Beijing and other large Chinese cities. In an article published on April 14, 2014 by The Wall Street Journal – China, the count of bad days in Beijing was finally available. The U.S. Embassy had been monitoring air quality in their compound in Beijing since 2008. The results are shown in two histograms shown below.

_____

[41] Chak K. Chana, Xiaohong Yao. "Air pollution in mega cities in China". *Atmospheric Environment,* Volume 42, Issue 1, January 2008, Pages 1–42.

## How Many Bad Air Days in Beijing?

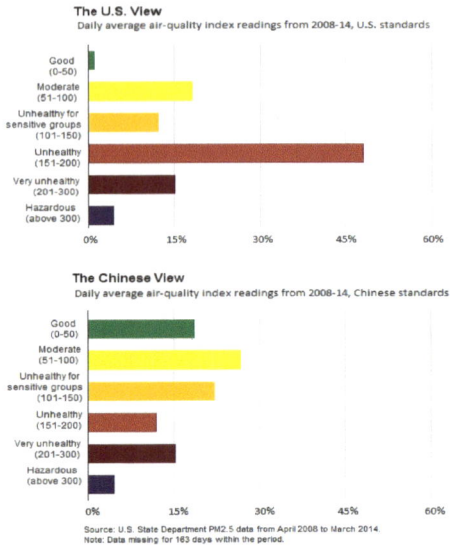

**The U.S. View**
Daily average air-quality index readings from 2008-14, U.S. standards

Good (0-50)
Moderate (51-100)
Unhealthy for sensitive groups (101-150)
Unhealthy (151-200)
Very unhealthy (201-300)
Hazardous (above 300)

0%   15%   30%   45%   60%

**The Chinese View**
Daily average air-quality index readings from 2008-14, Chinese standards

Good (0-50)
Moderate (51-100)
Unhealthy for sensitive groups (101-150)
Unhealthy (151-200)
Very unhealthy (201-300)
Hazardous (above 300)

0%   15%   30%   45%   60%

Source: U.S. State Department PM2.5 data from April 2008 to March 2014.
Note: Data missing for 163 days within the period.

FROM THE WALL STREET JOURNAL – CHINA. APRIL 14, 2014. RETRIEVED 06/22/2015 FROM HTTP://BLOGS.WSJ.COM/CHINAREALTIME/2014/04/14/BEIJINGS-BAD-AIR-DAYS-FINALLY-COUNTED/

"Based on data collected for 2,028 days between April 2008 and March 2014, only 25 days were considered "good" by U.S. standards."

In typical Chinese government fashion, the Olympic solution was to severely curtail local industrial activities for the period of the games. It is also quite possible that the particulate pollution may be so high that it is blinding Chinese officials who have prepared their own rendition of the data.

It must be remembered that this is U.S. Embassy data – a single monitoring station. In virtually any other situation this single-point data would be considered scientifically non-representative and, in any other situation I would have to agree, but we must accept this as a reality since the Chinese government will not accept the reality that their policies have created. So, the "pea-soupers" (London smog)[42] have become "bird's nest soupers" in Beijing.

There has been no hesitation or error in pointing out the 5 countries responsible for over half of the consumption of energy and production of $CO_2$ emissions. However, it is important not to simply stop here even though it would be statistically justified. If change is necessary the "devil is in the details".

The "details" mentioned above really are in the source of energy used for the particular sector or type of consumption. Although it should be very simple to present this data one must be exceptionally careful in comparing contexts and matching the qualifying conditions for data collection and presentation.

For most problems it is important to tackle the significant issues first, that is why the top five countries must remain the focus in the discussion of potential solutions or changes. The distribution of power consumption is critical to any discussion of measures that can reduce the global carbon footprint. No

---

[42] http://www.cleanerairforlondon.org.uk/londons-air/air-quality-data/trends-london/history-air-pollution-london. Retrieved 06/22/2015.

matter how well individual small countries use alternative fuels or trap their emissions the outcome will be irrelevant unless the major countries / consumers lead the way.

**EIA Beta ANALYSES - Energy Profiles - Top 10 Countries**

| DESCRIPTION / COUNTRY ---> | CHINA | USA | RUSSIA | INDIA | JAPAN | CANADA | BRAZIL | GERMANY | IRAN | S. KOREA |
|---|---|---|---|---|---|---|---|---|---|---|
| *Quadrillion BTUs* | | | | | | | | | | |
| Total Primary Coal Production: 2012 | 78.676 | 20.667 | 7.505 | 10.673 | 0 | 1.547 | 0.077 | 1.973 | 0.03 | 0.039 |
| Dry Natural Gas Production: 2012 | 3.831 | 24.61 | 22.33 | 1.497 | 0.2 | 5.242 | 0.633 | 0.436 | 5.965 | 0.042 |
| Total Primary Energy Production: 2012 | 101.781 | 79.119 | 55.296 | 15.874 | 1.568 | 19.139 | 9.758 | 4.804 | 13.644 | 1.558 |
| Production of Crude Oil including Lease Condensate: 2014 | 6.405 | 13.791 | 37.286 | 0.215 | | 4.738 | 1.856 | | 3.822 | 0 |
| *Quadrillion BTUs* | | | | | | | | | | |
| Total Coal Consumption: 2012 | 69.72 | 16.677 | 5.101 | 12.891 | 4.725 | 0.965 | 0.513 | 3.21 | 0.064 | 3.254 |
| Total Petroleum Consumption: 2013 | 21.052 | 35.099 | 6.807 | 7.079 | 9.145 | 4.732 | 6.132 | 4.959 | 3.816 | 4.77 |
| Dry Natural Gas Consumption: 2013 | 5.987 | 26.819 | 14.982 | 1.888 | 4.744 | 3.267 | 1.405 | 3.22 | | 2.106 |
| Total Primary Energy Consumption: 2012 | 105.882 | 95.058 | 31.522 | 23.916 | 20.306 | 13.354 | 12.095 | 13.466 | 9.645 | 11.52 |
| *% of world* | | | | | | | | | | |
| Total Electricity Installed Capacity: 2012 | 21.2 | | 4.2 | 4.6 | 5.3 | 2.4 | 2.2 | 3.2 | 1.4 | 1.7 |
| Proved Reserves of Natural Gas: 2015 | 2.5 | | 25.5 | 0.8 | 0 | 1.1 | 0.2 | 0.1 | 18.1 | 0 |
| Total Recoverable Coal: 2011 | 12.9 | | 17.7 | 6.8 | 0 | 0.7 | 0.7 | 4.6 | 0.1 | 0 |
| Crude Oil Proved Reserves: 2015 | 1.5 | | 4.9 | 0.4 | | 10.6 | 0.9 | 0 | 9.8 | |

DATA COMPILED FROM EIA BETA ENERGY ANALYSES RETRIEVED 06/2015 FROM
HTTP://WWW.EIA.GOV/BETA/INTERNATIONAL/COUNTRY.CFM?ISO= WHERE = IS FOLLOWED BY THE COUNTRY SHORT **ABBREVIATION.**

The highlighted section was not available in the same form, despite direct requests, from the very agency (EIA) that produces the Beta analyses. All US data had to be compiled by separate downloads from customized tabular EIA data or yearly reports.

It is very clear by looking at the above table that the Top 5 countries, both in terms of energy production and consumption, are by far the only significant players in the game of global climate change. A glimpse at Canada (number 6) shows that for 2012 Total Primary Energy Consumption was only 65.76% of Japan (number 5). This allows for streamlining a more detailed energy picture for the top 5 countries.

## TOP 5 POLLUTERS

**Energy for 2012[43]** *(in quadrillion BTU unless otherwise specified)*

| | CHINA | USA | RUSSIA | INDIA | JAPAN |
|---|---|---|---|---|---|
| ***Total Energy Production*** | *101.78* | *79.12* | *55.30* | *15.87* | *1.57* |
| Coal | 78.68 | 20.67 | 7.51 | 10.67 | 0.00 |
| Total Petroleum and Other Liquids | 8.79 | 13.79 | 21.35 | 1.63 | 0.01 |
| **Renewables** | **9.55** | **5.04** | **1.60** | **1.52** | **1.16** |
| Non-Hydroelectric Renewables | 1.40 | 2.41 | 0.03 | 0.34 | 0.45 |
| Geothermal | 0.00 | 0.15 | 0.01 | | 0.03 |
| Wind | 0.91 | 1.34 (s) | | 0.27 | 0.05 |
| Solar, Tide, Wave, Fuel Cell | 0.06 | 0.04 | | 0.02 | 0.07 |
| Solar | 0.06 | 0.04 | | 0.02 | 0.07 |
| Biomass and Waste | 0.43 | 0.88 | 0.03 | 0.05 | 0.32 |
| Hydroelectric | 8.15 | 2.63 | 1.57 | 1.19 | 0.71 |
| Nuclear | 0.94 | 8.06 | 1.83 | 0.36 | 0.18 |
| ***Total Energy Consumption*** | *105.88* | *95.06* | *31.52* | *23.92* | *20.31* |
| Coal | 69.72 | 16.68 | 5.10 | 12.89 | 4.73 |
| Petroleum (Oil & Gas) | 20.85 | 34.58 | 6.97 | 7.33 | 9.52 |
| Hydroelectric | | *All of these categories have negligible* | | | |
| Nuclear | | *import export numbers therefore* | | | |
| Non-Hydroelectric Renewables | | *Consumption ≈ Production* | | | |
| $CO_2$ Emissions (MM tons $CO_2$) | 8106 | 5270 | 1782 | 1831 | 1259 |
| **Renewables as % of Consumption** | **9.02%** | **5.30%** | **5.07%** | **6.36%** | **5.73%** |

Even a cursory view of this data permits a rapid determination that alternative or renewable energy, with the exception of nuclear, is generally insignificant for these countries or cannot be extensively increased[44]. The case for alternative or renewable energy is presented in its own chapter.

Undoubtedly the best summaries of a country's production and consumption of energy are produced by the US Department of Energy, Energy Information Agency (EIA). The Internet Beta analyses for many countries attempt to present comprehensive enough data in order to support predictive analyses.

---

[43] For various reasons, the latest year available for detailed comparison using DOE / EIA data is 2012. Some parameters can be compared for 2013.

[44] This is particularly the case for hydroelectric power where at least 4 of the five countries (the exception is India) have developed their large-scale hydro-electric potential to a very high degree and there is little other resource to tap without severe ecological consequence.

Unfortunately, the analysis for the United States does not permit a simple comparison due to its complexity and presentation by state and not as a country-wide aggregate.

Due to the nature of reporting and collecting information across international borders there must be some caution used in citing data as completely reliable. However, for the purposes of the arguments made here, the periodic inability to get numbers that sum completely to 100% is unimportant simply due to the fact that the overall energy distribution is really the key issue and the contributions of alternative or renewable energy to the production / consumption numbers is about 10 to 20 times less than fossil fuels.

## HOW CAN WE CHANGE THE SITUATION?

There are consequences to all actions and the decision to reign in $CO_2$ emissions has vital economic consequences for nations heavily tied to fossil fuel use. This is one of the essential reasons that change is not occurring and is highly unlikely to occur without some rather targeted measures. Some suggestions from the conventional to the extreme will be discussed later.

# SOLAR, WIND & OTHER ALTERNATIVE POWER

*"Solar, wind, hydroelectric and geothermal power are all the children of fusion – without the sun, none would exist."*

Duane Chartier (Recovering Scientist – Author of this book, 1952 – present)

It is sobering to get back to very basic concepts and notions. **Solar energy is fusion energy** delivered approximately 93,000,000 miles from its source. Without differential heating of the atmosphere and the associated differential movement of air there would be no wind. Geothermal energy is, to a large degree, the result of continued radioactivity keeping the core temperature of the planet elevated. Hydro-electric power is the release of the gravitational potential energy stored in water at a higher elevation than another. The water got to the higher levels by rain … weather – the heating of the earth by the sun.

Tidal forces are the only significant forces that are not directly solar but are, in fact, primarily lunar. So, to the purist or excessively scientific, almost every power source on earth is some-how related to the sun and thereby to fusion.

This entire point may seem to be labored but it is incredibly common to meet people who are as intensely "anti-nuclear" as they are "pro-solar". To a scientist this is a little bizarre and shows very little higher understanding of the relationships of energy and energy transfer on our planet, within our solar system and our universe.

## ALTERNATIVE / RENEWABLE ENERGY SOURCES

The EIA has a simple listing for renewable energy as cited below:

**"What is renewable energy?**

Unlike fossil fuels, which are finite, renewable energy sources regenerate.

There are five commonly used renewable energy sources:

Biomass—includes: Wood & wood waste; Municipal solid waste; Landfill gas & biogas; Ethanol; Biodiesel

Hydropower

Geothermal

Wind

Solar"[45]

# U.S. energy consumption by energy source, 2015

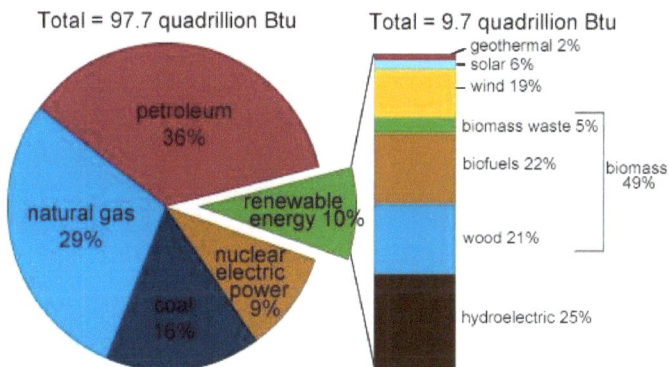

Total = 97.7 quadrillion Btu    Total = 9.7 quadrillion Btu

petroleum 36%
natural gas 29%
coal 16%
nuclear electric power 9%
renewable energy 10%

geothermal 2%
solar 6%
wind 19%
biomass waste 5%
biofuels 22%
wood 21%
biomass 49%
hydroelectric 25%

Note: Sum of components may not equal 100% because of independent rounding.

Source: U.S. Energy Information Administration, *Monthly Energy Review*. Table 1.3 and 10.1 (April 2016), preliminary data

eia

---

[45] http://www.eia.gov/energyexplained/index.cfm?page=renewable_home. Retrieved 11/18/2016.

This seems straightforward but there are several issues to be considered. In looking at the literature and Internet offerings on the subject of alternative energy one can go from the practical to the occult without any problem. The actual problem is in getting some realistic perspective as to the contribution of renewables or alternative energy sources to world production and consumption.

**As a small example of some of the conflation and confusion that can occur** all one has to do is consider the article on the "Top 10 Renewable Energy Sources"[46]. Although seemingly well researched, that provides the following flawed list:

1. Nuclear Power
2. Compressed Natural Gas
3. Biomass
4. Geothermal Power
5. Radiant Energy
6. Hydroelectricity
7. Wind Power
8. Solar Power
9. Wave Power
10. Tidal Power                    The yellow highlighting in the list indicates some problems:

Nuclear (#1) is listed as renewable source but this poses a real problem when comparing data or simply discussing an issue. Technically speaking Nuclear is NOT renewable if the source is fissionable material.

One totally non-renewable fossil fuel (Compressed Natural Gas #2) that they state is a fossil fuel but cleaner than the rest. How it made a renewable list is completely incomprehensible! Therefore, it is not discussed in this chapter.

Biomass (#3) is problematic because we are essential burning another material that is not fossil fuel but contributes $CO_2$ and therefore is a significant contributor to global warming.

The Radiant Energy section (#5) is not only entertaining but completely in left field, bordering on cultist[47], and very strange indeed, if the purpose of the authors had anything to do with actually informing anyone of anything useful.

---

[46] "Top 10 Renewable Energy Sources" by Listverse Staff, May 9, 2009. http://listverse.com/2009/05/01/top-10-renewable-energy-sources/. Retrieved 07/22/2015.

[47] The Methernitha Community in Switzerland apparently has "5 or 6 working models of fuelless, self-running devices that tap this energy" [Radiant Energy - source is in footnote 1]. The community website is http://www.methernitha.com/. Besides making energy machines they are "Eine christliche Gemeinschaft im Emmental" = "A Christian community in the Emmental". They only accept visitors by appointment and when I asked to visit to see the experiments and make measurements they were very polite in saying they no longer had anything to do with this but if I had Christian ideas I could visit for three days. Website retrieved 07/25/2015.

The list also does not present the power sources in any discernable logical order. Despite and because of this, the presentation of the sources below will start with tidal forces that are mostly non-solar related and will basically follow the list backwards with the omission of compressed natural gas (fossil fuel) and radiant energy.

## TIDAL POWER

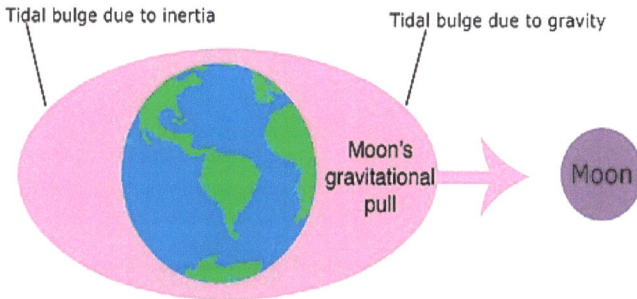

*Tidal bulge due to inertia*      *Tidal bulge due to gravity*

*Moon's gravitational pull*

*Moon*

Tidal forces are caused by the gravitational effects of the moon (and to a lesser degree, the sun) on the oceans as it orbits the earth. It is really not as simple as that because as the moon pulls water toward it there is a bugle created in the opposite direction caused by inertia. Of

*SIMPLE DIAGRAM RETRIEVED 07/01/2015 FROM NOAA'S OCEAN SERVICE EDUCATION (HTTP://OCEANSERVICE.NOAA.GOV/EDUCATION/KITS/TIDES/MEDIA/SUPP_TIDE03.HTML)*

course, this is still too simple in that there are higher tides when the sun and moon are aligned than when the moon is 90° from the sun.

Theoretically, there should be 2 high tides and 2 low tides per day. Unfortunately, this is still too simple because the continents effect the movement of the water as does the direction of the earth's rotation and this leads to more complex tidal cycles. NOAA (the National Oceanic and Atmospheric Administration - http://www.noaa.gov/) has good basic educational resources to study climate and the oceans in addition to exceptional monitoring and forecasting services and that those will provide a much fuller understanding of tidal patterns.

The essential points to be made about tides and their value for power generation are:

1. Tides vary considerably around the globe.
2. Tides are *local* in the sense that they are coastal in terms of ease of human access or developmental use.
3. Although the movement of water is massive it is essentially quite slow so generating power from the tides is easier said than done.
4. Tides are more predictable and reliable than wind energy and solar power.

There are two basic ways to exploit tidal energy – underwater tidal stream generators (propellers or turbines) or by a barrage[48], a dam-like structure, in a tidal basin or river mouth. Both methods are problematic and the claims of several companies seeking investment money for tidal power generation have fallen markedly short. A British company, Lunar Energy, announced that they would be building the world's first tidal energy farm off the coast of Pembrokshire in Wales. It was to be the world's first deep-sea tidal-energy farm and would provide electricity for 5,000 homes. Eight underwater turbines, each 25 metres long and 15 metres high, were to be installed on the sea bottom off St David's peninsula. Construction was supposed to have started in the summer of 2008 and the "wind farm under the sea", was scheduled to be operational by 2010.[49] This has not materialized as promised.

The Rance Tidal Power Station is a tidal power station (a barrage about 750 m (2,461 ft.) long) located on the estuary of the Rance River in Brittany, France. It was opened in 1966 as the world's first tidal power station. It is operated by Électricité de France and was for 45 years the largest tidal power station in the world by installed capacity. Its 24 turbines reach peak output at 240 megawatts and average 62 megawatts, a capacity factor of approximately 26% with annual output of approximately 540 GWh, it supplies 0.012% of the power demand of France.[50] Regardless of the commendable notion and effort no one would ever consider this anything but an insignificant or token nod to renewable energy.

Another example is the large (1,000-tonne, five-story) turbine used to glean power from the largest tidal bore in the world – on the floor of the Minas Passage in the Bay of Fundy in Nova Scotia Canada. The turbine was lowered into place in early November 2016 and was rapidly grid-connected. It is capable of generating two megawatts of power. Although this is a commendable achievement it does come at a very high price. "The electricity being generated is some of the most expensive ever produced in Nova Scotia, costing $530 per megawatt hour versus the current average of $60 per megawatt hour."[51]

This high cost will eventually come down as the technology becomes better but, the present very high cost is being subsidized by utility users and a Canadian government program (taxpayers). It will take many years to make a proper real-world assessment of the success of the project despite the excitement at being a first in North America.

---

[48] A barrage is an obstruction in a watercourse, usually to increase the height of the water but, in this case, to house the turbines to collect the tidal energy.

[49] http://listverse.com/2009/05/01/top-10-renewable-energy-sources/. Retrieved 07/22/2015. Checked 02/21/2017.

[50] https://en.wikipedia.org/wiki/Rance_Tidal_Power_Station. Retrieved 07/20/2015.

[51] CBC News – Nova Scotia. Initially reported November 22, 2016. Updated November 24, 2016. http://www.cbc.ca/news/canada/nova-scotia/tidal-power-bay-of-fundy-turbine-electricity-emera-hydro-1.3862227. Retrieved 06/18/2017.

## OCEAN CURRENTS

Although not listed directly in the opening list, ocean currents are tied together with tides and other effects (bad pun again intended).  As NOAA indicates:

> "Ocean currents can be generated by wind, density differences in water masses caused by temperature and salinity variations, gravity, and events such as earthquakes.
>
> Surface currents are generated largely by wind.  Their patterns are determined by wind direction, Coriolis forces from the Earth's rotation, and the position of landforms that interact with the currents.  Surface wind-driven currents generate upwelling currents in conjunction with landforms, creating deep water currents.  Currents may also be generated by density differences in water masses caused by temperature and salinity variations.  These currents move water masses through the deep ocean—taking nutrients, oxygen, and heat with them."[52]

Currents can be very strong in particular situations such as near undersea mountains and canyons. There are some very strong and predictable currents such as the Antarctic Circumpolar Current[53] that moves perpetually eastward under the driving force of the westerly "roaring 40's" winds.  The current transports 130 million cubic metres per second (4.6×109 cu ft/s) of water – 100 times the flow of all the world's rivers.  However, harnessing the power of such a current and transmitting the energy to where it can be effectively used is a significant issue and is presently a barrier to development.

## WAVE POWER

The kinetic energy of waves is a direct result of wind turbulence at the water surface.  As with all forms of renewable energy there are pros and cons but an honest assessment of the viability of using waves must be more negative than positive, waves are highly variable in location and intensity.  The technology to harness wave motion is not straightforward and is sometimes down-right strange.  For example, wave farms have been created and are in use in Europe, using floating Pelamis Wave Energy converters.

There are several systems for harnessing wave energy.  A good cross-section of technology is presented (in brief) by the US Bureau of Ocean Energy Management (BOEM) on their website[54] under the section for "Ocean Wave Energy".  Although the potential for wave energy is high in specific coastal areas there

[52] http://oceanexplorer.noaa.gov/facts/currents.html.  Retrieved 07/20/2015.  I have no problem using passages from any source that are better written than I would have produced.
[53] https://en.wikipedia.org/wiki/Southern_Ocean.  This is the largest and strongest current.  Retrieved 07/25/201.
[54] https://www.boem.gov/Renewable-Energy-Program/Renewable-Energy-Guide/Ocean-Wave-Energy.aspx. Retrieved 06/18/2017.

have only been small-scale projects and the entire approach seems to be at the experimental and trial stage level.

It is possible that the potential marine hazards and capital costs are highly unlikely to be offset by the actual power produced. The amount of power produced from wave action is still negligible[55] in a global discussion of power.

## SOLAR POWER

Solar power is the production of electricity from solar radiation and is fundamentally divided into active (direct) and passive (indirect) production. Active production is through photovoltaic (PV) devices[56] and passive production is generally done by using concentrated solar power (CSP).

The PV devices create electricity directly. Initially they were small, expensive and did not produce large amounts of power. For this reason, they were used more in novelty or convenience items or those that did not require much power such as a solar powered calculator.

There are now some large "PV farms"[57] such as the 579 Mw Solar Star facility[58] which is co-located in Kern and Los Angeles counties in southern California. At present, it is the largest PV farm in the world.

The installations that use CSP technology tend to be more intrusive in that solar energy has to be focused onto collectors either in a distributed array or into a centralized collector to create the elevated temperatures to produce the steam that runs the turbine / generator that, in turn, produces the electricity. Essentially this is like any thermal power plant that converts heat into electricity.

There are three basic types: parabolic trough, tower and parabolic dish. There is an additional type that is a parabolic dish collector that does not heat steam but creates electricity through a Stirling Engine[59].

---

[55] As a scientist I like to quantify even things like negligible – for these discussions it means about 0.08% - about the blood alcohol level that will just get you into trouble on a breathalyzer test "0.08 BAC is legally impaired and it is illegal to drive at this level" (per http://www.lifeloc.com/measurement.aspx retrieved 11/17/2015).

[56] Photovoltaics are solid state devices that produce electric current when exposed to sufficient solar radiation.

[57] Author's term.

[58] The fact sheet is available from http://us.sunpower.com/sites/sunpower/files/media-library/fact-sheets/fs-solar-star-projects-factsheet.pdf. Retrieved 07/27/2015.

Trough systems use large, parabolic reflectors or focusing mirrors that have pipes filled with oil or other heat transfer agent running along their center, or focal point. The mirrored reflectors are tilted toward the sun, and focus sunlight on the pipes to heat the fluid inside to as much as 750°F. The hot fluid is then used to boil water, which makes steam to run conventional steam turbines and generators.

The power tower is a central collector that receives reflected sunlight from tracking mirrors or heliostats.

Parabolic dish collectors are like a small-scale tower operation and each dish has its own collector.

It is difficult to assess the credibility of critiques of solar installations. Certainly, there is a tendency of any entity seeking investors to overstate the return on investment either from overconfidence or failure to be realistic – i.e. using the most optimistic projections of degree-days, maintenance costs and other expenditures. One of the inherent problems with any solar or wind installation is that the power density is not guaranteed so that all reasonable production projections should be using statistically accurate meteorological parameters for the specific site. This data is always best prepared by third party experts with no vested interest in the outcome … Wow! A proposal to do legitimate scientific and technical assessments – many of which have already be done in earnest.

Because of the increasing importance and prevalence of solar power projects it is necessary to scale the relative contributions to net production (or consumption) of energy.

## WIND

Although winds are intrinsically variable there are geographic locales where they are fairly reliable. Historically, the windmills of the Low Countries (a coastal region of western Europe, consisting mostly of the Netherlands and Belgium, and the low-lying delta of the Rhine, Meuse, Scheldt, and Ems rivers where much of the land is at or below sea level), particularly Holland (now the Netherlands) represented the first systematic and widespread use of wind power. The Dutch windmills were used to pump water from the polders[60]. There a number of quite interesting historical accounts of harnessing wind energy but that would require another book and detract from the purpose here.

---

[59] http://solarcellcentral.com/stirling_page.html. Retrieved 07/27/2015.
[60] Polders are areas of low-lying land reclaimed from the sea or a river and protected by dikes.

The historical interest in wind power has continued in the modern Netherlands but not to any significant degree according to analysis by the EIA:

> "The Netherlands net electricity generation was approximately 94.9 billion kilowatthours (BkWh) in 2012, mostly from fossil fuel-fired power plants. More than 12% of the Netherland's electricity is generated from renewable sources, mainly biomass and waste (7% of the total) and wind (5% of the total)."[61]

The Netherlands, one of the countries in the world with the highest percentage production of wind power, is still only at 5% of total power production. **This is not promising for the world.**

Although windmills and wind farms are a popular environmental topic and the growth of wind power has been impressive world-wide there are still some fundamental problems with the source itself and with assessing (objectively) the overall importance of wind power. The bulleted items summarize some of the concerns and provide the rationale for moving on to other issues:

- There are fundamental geographic and meteorological constraints to the siting of wind fields to insure the highest return on investment. Many areas are simply not appropriate to provide sufficient wind power – at least at an industrial level.
- The full range of environmental effects has not been completely or objectively evaluated.
- There is not a sufficient or reliable database of capital, maintenance and operating costs to be able to perform a truly objective assessment of efficacy and a comparison to other renewable power production.

## HYDROELECTRICITY

Hydroelectricity has its roots deep in the past as dams have been used for millennia to trap water and use its gravitational potential energy. Initially this was all mechanical energy to run mills or pumps or tools by running the water over a water wheel and then over more carefully designed turbines to effect rotational motion in a shaft. As with wind power, the transition to produce electricity directly came with the advent of efficient generators.

Hydroelectricity is indeed renewable but the costs are often highly underrated or ignored. Dams cause fundamental and often large-scale changes to the environment both upstream and downstream. The largest dam project built to date is the Three Gorges Dam that spans the Yangtze River near the town of Sandouping. The Three Gorges Dam is the world's largest power station in terms of installed capacity

---

[61] http://www.eia.gov/beta/international/analysis.cfm?iso=NLD. Retrieved 08/14/2015.

(22,500 MW).  The dam is the largest operating hydroelectric facility in terms of annual energy generation of 98.8 TWh in 2014[62] compared to the Itaipú Dam in Brazil / Paraguay at 87.8 Twh in 2014[63].

Although the Chinese government regards the project as a historic engineering, social and economic success[64], creation of the dam flooded archaeological and cultural sites and displaced approximately 1.3 million people, and is causing significant ecological changes, including an increased risk of landslides[65].  The dam has created considerable cultural disruption and has been the target of some internal Chinese criticism but there is, of course, far more from international critics.

Aside from potential large scale ecological changes upstream, the downstream reduction of water flow can be significant.  The dams along the Colorado River provide a graphic example of extreme ecological changes downstream.  The simple fact remains that once the changes (some would say damages) have been done there is really no way to turn back without removing the dam.

Another inherent limitation of hydroelectric power is location.  The sites for developing hydroelectric potential are completely dictated by geography and geology.  Developing large scale hydroelectric power projects any extended distance from where the power will actually be used requires large-scale investment and a development of a power grid capable of distribution at large distances.  There is a particularly good example in The James Bay Project.  Hydro-Québec constructed a series of hydroelectric power stations on the La Grande River in northwestern Quebec, Canada, and the diversion of neighboring rivers into the La Grande watershed.  The project covers an area of the size of the State of New York and is one of the largest hydroelectric systems in the world.  It has cost up to US$20 billion to build and installed generating capacity of 16,527 megawatts.  "If fully expanded to include all of the original planned dams, as well as the additional "James Bay II" projects, the system would generate a total of 27,000 MW [citation needed] making it the largest hydroelectric system in the world."[66]

The region is inhabited by Cree and Inuit First Nations peoples and is a vast area.  The La Grande River watershed stretches over 177,000 km² (68,000 sq mi) or approximately 11% of the total area of Quebec, an area larger than the state of Florida or twice the size of Scotland.[67]  There was a very strong and effective backlash with approximately 5000 Cree over land rights and environmental issues and the

---

[62] "Three Gorges breaks world record for hydropower generation".  Xinhua. 1 January 2014.  Retrieved 10/11/2015.

[63] "Generation". Itaipu Binacional. Retrieved 10/11/2015.

[64] "中国长江三峡工程开发总公司".  Ctgpc.com.cn. 2009-04-08.  Retrieved 10/11/2015.

[65] "重庆云阳长江右岸现 360 万方滑坡险情-地方-人民网".  *People's Daily*.  Retrieved 10/11/2015.  See also: "探访三峡库区云阳故陵滑坡险情".  *www.News.xinhuanet.com*.  Retrieved 10/11/2015 (dated 2009-04-09.)

[66] https://en.wikipedia.org/wiki/James_Bay_Project.  Retrieved 10/11/2015.

[67] Ibid.

Quebec government was forced to come to a settlement. This has substantially altered the original grand scheme that would have made it the largest system in the world.

"The Agreement Respecting a New Relationship Between the Cree Nation and the Government of Quebec (dubbed as La Paix des Braves, French for "The Peace of the Braves" by the Parti Québécois government) is an agreement between the Government of Quebec, Canada, and the Grand Council of the Crees. It was signed on February 7, 2002 in Waskaganish, Jamésie, Quebec, after decades of court battles between the Cree and the Government of Quebec. The name was inspired by the 1701 Great Peace of Montreal, also known as "La Paix des Braves"."[68]

There is considerable background information on hydroelectric power development that is deliberately not presented here. However, the abstracted essential points to be made are:

- most developed nations have developed their large-scale hydroelectric potential to the maximum. There is still room for some additional power with "low-head" hydro.[69] However, this has limited potential and there are extended environmental changes that must be assessed systematically
- hydroelectric projects (especially large ones) often grossly underestimate environmental and other less tangible effects such as cultural / lifestyle change and general quality of life issues
- The 5 major polluters, who need to reduce their carbon footprints drastically, are not positioned to use hydroelectric development as a tool to mitigate their energy use.

## GEOTHERMAL

There is a large geothermal potential energy as a result of general geological activity. However, the problem is access and surface availability and the unfortunate irony is that geologically active zones where geothermal energy is available are inherently unstable. Another determinant is location relative to potential users. Basically, anyplace on the "ring of fire"[70] has the potential to produce significant geothermal energy. In addition, there are hot spots such as old faithful in Wyoming and the Hawaiian Islands that are actually rips in the continental or ocean plate.

---

[68]https://en.wikipedia.org/wiki/Agreement_Respecting_a_New_Relationship_Between_the_Cree_Nation_and_the_Government_of_Quebec. Retrieved 06/10/2017.

[69] https://en.wikipedia.org/wiki/Low_head_hydro_power. Retrieved 06/21/2017.

[70] "The **Ring of Fire** is a major area in the basin of the Pacific Ocean where a large number of earthquakes and volcanic eruptions occur. In a 40,000 km (25,000 mi) horseshoe shape, it is associated with a nearly continuous series of oceanic trenches, volcanic arcs, and volcanic belts and/or plate movements." https://en.wikipedia.org/wiki/Ring_of_Fire. Retrieved 07/04/2017.

The largest group of geothermal power plants in the world is located at The Geysers, about 72 miles north of San Francisco, California. The Geysers is the world's largest geothermal field, containing a complex of 22 geothermal power plants, drawing dry superheated steam from more than 350 wells. This is a somewhat unusual location and the heat source is a magma chamber that is relatively close to the surface. It is a relatively stable seismic area but there are indications that seismic activity may be increasing with one possibility that the increase may be related to drilling the wells and changing release pressures. It is clear that the large-scale development of the site is related to its proximity to major population centers – the users.

As of 2004, five countries (El Salvador, Kenya, the Philippines, Iceland, and Costa Rica) generate more than 15% of their electricity from geothermal sources.[71]

## BIOMASS

The use of biomass to generate power seems to be a reasonable approach to the general waste management problem and renewability. However, there are many inherent problems that have serious climatic implications. How can burning any organic material to produce net $CO_2$ be a good thing? So, biomass is renewable because … we have, for all-intents-and purposes, grossly mismanaged our excessive consumption of energy and waste production. Many people also seem to think that growing crops such as corn (for ethanol) or trees to make burnable fuel pellets can produce a quasi-infinite supply of materials that we will then burn.

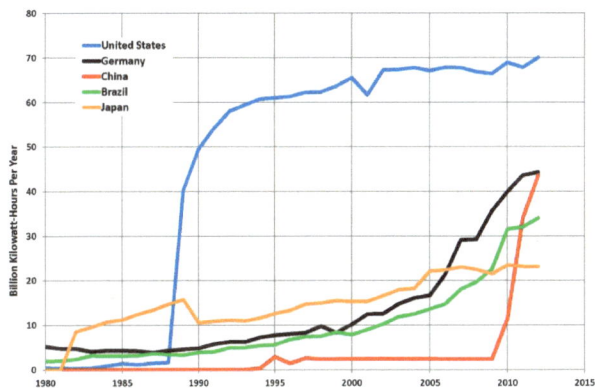

**This is the world upside down! Let's grow plants to "green the planet" to destroy the climate which will, in turn, destroy the planet.**

This graphic indicates biomass energy production.

HTTPS://EN.WIKIPEDIA.ORG/WIKI/BIOMASS. RETRIEVED 10/11/2015.

---

[71] https://en.wikipedia.org/wiki/Geothermal_power. **Retrieved 06/21//2017.**

It should be no surprise that there is a rise in Japanese biomass energy production after the Fukushima nuclear disaster and the incredible decision to close all of their nuclear plants. However, it is relatively small given the very steep rise in biomass energy production in China from about 2009. Germany's biomass energy production has also strongly increased since about 2000 and Brazil's has climbed steadily upward. Although US biomass energy is nearly double any other country the curve has levelled off.

## NUCLEAR

**If nuclear power were only fusion power, then it would be effectively limitless and renewable.** However, as already discussed, there is no commercial power produced by fusion and it is a minimum of a decade away unless fundamental thinking at a government or international level changes and significant increases in R&D money is expended.

All of the nuclear power in the world is produced by fission. **In the strictest sense power produced by nuclear fission should not be considered renewable** in that there is a limited fissionable material supply available. The only possible rationale for considering nuclear as renewable is that the energy yield from fissionable material is enormous relative to the any other fuel source. This entire, seemingly semantic, problem underlines a much deeper problem and that is the inability, or worse, the refusal to think and communicate clearly on what problems actually exist. It is almost inconceivable that there has been very little serious effort by the US Department of Energy to clearly and decisively point out any of the anomalies and contradictions in definitions, direction and policy.

Nuclear power production by fission is discussed in other parts of this book and there is no reason to place it in the wrong place, under renewable energy, to follow suite with what has been previously done. However, one of the major points of this book is that we need to pursue fission power at an accelerate rate due to the severity of the climate crisis.

# GLOBAL PERSPECTIVE - RENEWABLE / ALTERNATIVE ENERGY

Eco Watch[72] provided the following very promising headline "5 Countries Leading the Way Toward 100% Renewable Energy". The following major points were abstracted from the overall short article.

**"1. Denmark sets world record for wind** in 2014, getting 39.1 percent of its overall electricity.

**2. UK wind power smashes annual records in 2014.** A combination of grid-connected wind farms and standalone turbines produced 9.3 percent of the UK's electricity demand in 2014.

**3. Renewables provide biggest contribution to Germany's electricity.** Renewable energy was the biggest contributor to Germany's electricity supply in 2014, with nearly 26 percent of the country's power generation coming from clean sources.

**4. Scotland sees "massive year" for renewables.** December, 2014 was a "massive year" for renewables in Scotland. Over six months of the year, wind generated enough power to supply more than 100 percent of Scottish households, while in Aberdeen, Edinburgh, Glasgow and Inverness there was enough sunshine to provide 100 percent or more of the electricity needs for an average home in June and July.

With figures like these it is no wonder new research out this week said the country's power grid could be 100 percent renewable by 2030.

**5. Ireland hits new record for wind energy."**

## This would be incredible news and very important if it were not so unimportant!

The unfortunate reason that it is not important is the overall context. How would making all of these countries 100% powered by renewable energy have any significant impact on global $CO_2$ levels and global warming? **It would not** – the countries are small contributors.

A problem from the outset is the reporting of UK, Ireland and Scotland as essentially different since they are not in terms of most statistical representations of energy production and consumption. They are all essentially the UK. So really, the reporting is for 3 entities: Denmark, Germany, and the UK.

|  | World Ranking[73] | | |
|---|---|---|---|
|  | *DENMARK* | *U.K.* | *GERMANY* |
| Total Primary Energy Consumption 2012 | 67 | 13 | 6 |
| Total Primary Energy Production 2012 | 57 | 24 | 25 |

[72] http://ecowatch.com/2015/01/09/countries-leading-wat-renewable-energy/. Retrieved 10/08/2015.
[73] http://www.eia.gov/beta/international/. Reformatted and retrieved 10/11/2015.

Germany is the only one of the three countries that is a significant consumer of energy relative to the top five polluting countries and its nominal strides in conversion to renewable electricity is particularly irrelevant given its very high level of imported energy (3$^{rd}$ in the world, behind Japan, and only slightly behind the US in importation of dry natural gas in 2012).

In the "purported race" towards renewable energy the leaders would undoubtedly be China and the United States who invest the most money in renewable energy.

In another internet headline "Renewables Re-energized: Green Energy Investments Worldwide Surge 17% to $270 Billion in 2014". Tue, Mar 31, 2015 - See more at: http://www.unep.org/newscentre/default.aspx?DocumentID=26788&ArticleID=34875#sthash.dwOxlHu8.dpuf"

"China saw by far the biggest renewable energy investments in 2014 — a record $83.3 billion, up 39% from 2013. The US was second at $38.3 billion, up 7% on the year but well below its all-time high reached in 2011. Third came Japan, at $35.7 billion, 10% higher than in 2013 and its biggest total ever."[74]

**This race really takes more the form of a farce than anything else**. In general, any race has a specific goal. In the case of the major consuming countries that have invested the most in renewables only the most naïve would contend that the goal had anything to do with environmentalism. It is, most simply, an extension of the overall ramping of consumption of energy. Real concern for global climate problems would be indicated *by sustained and significant relative growth of renewable energy* use rather than a simple scaling of an absolute amount of money spent or power produced. **This numbers game played by the energy agencies of countries and their politicians is a very cynical and dangerous one to play.**

One productive way to look at renewable energy is to look at the percentage of the total energy consumption that it represents. Another method is to look at the **relative** rate of increase in renewable energy production. However, all of the tabular comparisons are not important in trying to solve the global energy crisis by using renewables such as solar and wind. This is by no means suggesting that these avenues should not be aggressively pursued by everyone on the planet. The real problem is that they are not being pursued with any honesty or credibility by the 5 major offenders.

---

[74] http://www.unep.org/newscentre/default.aspx?DocumentID=26788&ArticleID=34875. Retrieved 10/11/2015.

It is clear from the table below that the China, the USA and India all ramped down spending on renewable energy in 2013, but by a small percentage, so this does not yet constitute a firm indicator of policy.

**Table 3:** Global Investment in Renewable Energy by Region, 2004 – 2014

| | | 2004 | 2005 | 2006 | 2007 | 2008 | 2009 | 2010 | 2011 | 2012 | 2013 |
|---|---|---|---|---|---|---|---|---|---|---|---|
| USA | billion USD | 5.7 | 11.9 | 28.2 | 34.5 | 36.2 | 23.2 | 34.7 | 53.4 | 39.7 | 35.8 |
| America (excl USA and Brazil) | billion USD | 1.4 | 3.4 | 3.4 | 5.0 | 5.6 | 5.9 | 11.5 | 8.7 | 9.9 | 12.4 |
| Brazil | billion USD | 0.5 | 2.2 | 4.2 | 10.3 | 12.5 | 7.9 | 7.7 | 9.7 | 6.8 | 3.1 |
| Middle East and Africa | billion USD | 0.6 | 0.6 | 1.2 | 1.7 | 2.7 | 1.7 | 4.3 | 3.2 | 10.4 | 9.0 |
| Europe | billion USD | 19.6 | 29.4 | 38.4 | 61.7 | 72.9 | 74.7 | 102 | 115 | 86.4 | 48.4 |
| India | billion USD | 2.4 | 3.2 | 5.5 | 6.3 | 5.2 | 4.4 | 8.7 | 12.6 | 7.2 | 6.1 |
| China | billion USD | 2.6 | 5.8 | 10.2 | 15.8 | 25.0 | 37.2 | 36.7 | 51.9 | 59.6 | 56.3 |
| Asia and Oceania (excl. India and China) | billion USD | 6.7 | 8.3 | 8.9 | 11.0 | 11.5 | 13.2 | 20.7 | 25.3 | 29.5 | 43.3 |
| **Total** | billion USD | **39.5** | **64.8** | **100** | **146.3** | **171.6** | **168.2** | **226.7** | **279.6** | **249.5** | **214.4** |

Note: Data include government and corporate R&D.                    Source: BNEF and UNEP. Reference see endnotes

## PERSPECTIVE ON PERCEPTION

One of the singular most alarming pieces of data that I, as an author and scientist, have to process is the enormous gap between public perceptions and the actual measures that need to be taken to combat global climate change. *It is a very telling and disturbing tale when coal beats nuclear by 10% in the public view.*

## Global public support for energy sources

"Please indicate whether you strongly support, somewhat support, somewhat oppose, or strongly oppose each way of producing energy"

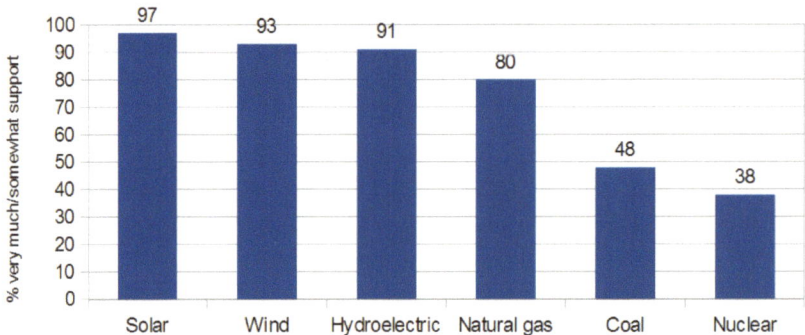

Source: Ipsos, May 2011

This single graphic was one of the factors that made me stop writing this book for at least 8 months. **I could not get my head around how totally inane this ordering is for a variety of very sound fundamental reasons.**

There is NO STATISTICAL OR FACTUAL DOUBT that coal mining and burning of coal has killed vastly larger numbers of people than nuclear power production and uranium mining. In a relevant article "Coal: The World's Deadliest Source Of Energy" by Nick Cunningham (May 14, 2014, 5:12 PM CDT)[75] one estimate of **annual** worldwide deaths from coal extraction was a very high 12,000. Even if the number were the result of a 10-fold error it would still be 1200/year. In the 1970's U.S. annual deaths were reported to be 141 on average. This is a staggeringly large loss and does not address the pollution related deaths due to consumption / burning of coal.

---

[75] http://oilprice.com/Energy/Coal/Coal-The-Worlds-Deadliest-Source-Of-Energy.html. Retrieved 11/29/2016.

There is a complete operational and logical disconnect between what is perceived to be a problem and what really is a problem. **The deaths and environmental damage from nuclear power mining and production are INSIGNIFCANT when compared to coal at virtually any level and at any time!**

This represents an enormous task for anyone who must or should take it on – convincing people that what they believe is not correct and educating them to come to statistically and logically based conclusions rather than blind acts of faith, gut reactions, and appeals to supposed data that really does not exist or has been invented.

As a former scientist who had to teach students something, I was never incredibly interested in their "impressions" just in what they actually knew! That may summarize where a graphic like the one above is daunting. How does one reverse notions that are simply notions and not based on any objective reality or even a mutually agreed upon fantasy?

"Aye, there's the rub!"[76] … the difficulty in the intellectual terrain that impedes forward progress. What is meant by this statement is layered and complex. In democratic countries, there are no restrictions on spouting utter nonsense or in appealing to peoples' worst fears. The NUCLEAR power industry employs very few people and has caused minimal loss of life and negligible environmental damage but people are incredibly concerned about nuclear safety. Coal mining and burning employs many more people world-wide and demonstrably kills many more people and dramatically and negatively alters the environment but there seems to be no significant concern at all. This is completely topsy-turvy.

Hopefully a more optimistic and operationally sound approach to dealing with this problem will arise in the discussion of solutions to climate change.

---

[76] (Hamlet, 3:1): "To sleep, perchance to dream—ay, there's the rub,
For in that sleep of death what dreams may come … Must give us pause "

# ENERGY POLICY OR RATHER, ENERGY FOLLY

*"We cannot solve our problems with the same thinking we used when we created them."*

Albert Einstein (German Physicist, March 14, 1879 - April 18, 1955)


If there is to be change - and there must be - then old habits (old ways of thinking) must be put to rest. Is this easy? NO. Is it absolutely necessary? YES. First it would be wise to have a list of "old habits" to address. Then it may be possible to discuss the present "energy policy" in the United States and to expand that to other relevant nations.


## OLD HABITS

Perhaps it would be more politically correct to select a different name for thinking and behaving in ways that are outmoded or inappropriate for solving modern problems. Unfortunately, Old Habits will have to do.

1. The NIMBY syndrome (**N**ot **I**n **M**y **B**ack **Y**ard) is alarmingly prevalent in local as well as international affairs. The basic translation of this approach is that although we may know something is intrinsically not a good practice it is done anyway because the local effects and consequences have been exported and/or diluted. A good example is the banning of most usage of the pesticide DDT in the United States in 1972 by the EPA but the continued sale of it and its use in many other countries.
2. The premise that countries need to be in competition for resources rather than co-developing them.
3. Confusing the difference between legitimate ideological differences and enmity.
4. The enemy of my enemy is my friend. An operationally flawed foreign relations and personal approach to dealing with interactions between interest groups and political forces.
5. Assuming that elected officials are actually working in the general interests of the public or for the public good.
6. Politicians who assume that people are too stupid to remember the last lies that they have been told! ... and the people who accept this behavior.


A brave attempt will be made to decipher the energy policies of the world's five largest consumers of energy and producers of $CO_2$ – China, United States, India, Russia, and Japan. Before diving directly into energy policy, it is valuable to consider the actual energy reserves of the countries as well as their production and consumption. The approach taken in presenting this data is that of a scientist. The five largest polluters account for over 50% of global $CO_2$ emission and their percentages of energy consumption are commensurate. Therefore, any real solution must come from these 5 countries and that is where the analysis and comments are directed.

### Fossil Fuel Reserves - Top 5 Polluters

| Country | Coal (2006) million tonnes | World Rank | World (%) | Oil (start 2015) millons of bbl | World Rank | World (%) | Natural gas proven reserves (m³) | World Rank | World (%) |
|---|---|---|---|---|---|---|---|---|---|
| *World* | *909,064* | --- | *100* | *1,662,945* | --- | *100* | *187,300,000,000,000* | --- | *100* |
| United States | 246,643 | 1 | 27.13 | 39,933 | 10 | 2.40 | 9,860,000,000,000 | 5 | 5.26 |
| Russia | 157,010 | 2 | 17.27 | 80,000 | 8 | 4.81 | 48,700,000,000,000 | 1 | 26.00 |
| China | 114,500 | 3 | 12.60 | 24,649 | 14 | 1.48 | 4,643,000,000,000 | 10 | 2.48 |
| India | 92,445 | 4 | 10.17 | 5,675 | 21 | 0.34 | 4,232,000,000,000 | 27 | 2.26 |
| Japan | ~0 | N/A | 0 | 44.1 | 78 | 0.00 | 20,900,000,000 | 77 | 0.01 |
| | | | 67.17 | | | 9.04 | | | 36.01 |

**Coal** - "BP Statistical review of world energy June 2007". British Petroleum. June 2007. Retrieved 2007-10-22. Wiki [4] ref.
**Oil** - U.S. Energy Information Administration, International Energy Statistics, accessed 3 Sept. 2016. Wiki {1} ref. bbl = barrel.
**Natural Gas** - https://en.wikipedia.org/wiki/List_of_countries_by_natural_gas_proven_reserves. References 2010-16.

### Fossil Fuel Production - Top 5 Polluters

| Country | Coal Mil. Tonnes (Mt) | World Rank | World (%) | Crude Oil Mil. tonnes (Mt) | World Rank | World (%) | Natural gas (Bm³) | World Rank | World (%) |
|---|---|---|---|---|---|---|---|---|---|
| *World* | *7,686* | --- | *100* | *4,296* | --- | *100* | *3,592* | --- | *100* |
| China | 3,538 | 1 | 46.03 | 216 | 4 | 5.03 | 137 | 5 | 3.81 |
| United States | 820 | 2 | 0.09 | 555 | 2 | 12.92 | 769 | 1 | 21.41 |
| India | 764 | 3 | 0.08 | 43 | 18 | 1.00 | 32 | 22 | 0.89 |
| Russia | 349 | 6 | 0.04 | 532 | 3 | 12.38 | 650 | 2 | 18.10 |
| Japan | 0 | 38 | 0 | 0 | 37 | 0.00 | 3 | 33 | 0.08 |
| | | | 46.24 | | | 31.33 | | | 44.29 |

**Coal** - Enerdata "Global Energy Statistical Yearbook 2016". Data is effectively 2015.
**Oil** - Enerdata "Global Energy Statistical Yearbook 2016". Data is effectively 2015.
**Natural Gas** - Enerdata "Global Energy Statistical Yearbook 2016". Data is effectively 2015.
https://www.enerdata.net/publications/world-energy-statistics-supply-and-demand.htm

### Fossil Fuel Consumption - Top 5 Polluters

| Country | Coal (2013) million tonnes | World Rank | World (%) | Oil (2015) (bbl/day) | World Rank | World (%) | Natural gas (m³) | World Rank | World (%) |
|---|---|---|---|---|---|---|---|---|---|
| *World* | *7,975* | --- | *100* | *92,430,923* | --- | *100* | *3,235,067,219,450* | --- | *100* |
| China | 3992 | 1 | 50.06 | 11,968,000 | 2 | 12.95 | 149,999,992,832 | 3 | 4.64 |
| United States | 837 | 2 | 10.50 | 19,396,000 | 1 | 20.98 | 689,900,027,904 | 1 | 21.33 |
| India | 808 | 3 | 10.13 | 4,159,000 | 3 | 4.50 | 64,490,000,384 | 10 | 1.99 |
| Russia | 207 | 5 | 2.60 | 3,113,000 | 7 | 3.37 | 457,200,009,216 | 2 | 14.13 |
| Japan | 190 | 7 | 2.38 | 4,150,000 | 4 | 4.49 | 112,599,998,464 | 5 | 3.48 |
| | | | 75.67 | | | 46.29 | | | 45.57 |

**Coal** - https://yearbook.enerdata.net/coal-and-lignite-world-consumption.html. Retrieved 02/02/2017
Total 2013 world coal consumption from IEA "Key Coal Trends 2016" p..
**Oil** - https://en.wikipedia.org/wiki/List_of_countries_by_oil_consumption. Retrieved 02/01/2017. Primary source IEA.
**Natural Gas** - http://www.indexmundi.com/g/r.aspx?t=0&v=137&l=en. Source: CIA World Factbook as of Jan 1, 2014.

N.B. The tables were compiled from a variety of data sources. The World Rank was done by simple data sort and the World % was added by the author.

Even the most cursory analysis of the data presented in the three tables above indicates that there is a strong similarity in the top four countries (China, United States, India, and Russia):

- They are large scale producers of the fossil fuel energy they are using. However, China's consumption has outstripped its production for several years so it has become a net importer of all fossil fuels.
- They have significant reserves of fossil fuels. India is the weakest in terms of reserves but still has considerable reserves and production capabilities.

Japan presents a completely different situation. Japan has very poor reserves and is a major importer of all fossil fuels. Its overall primary fossil fuel energy consumption far exceeds its capacity to produce that energy based on its own resources. Japan imported more natural gas than China, United States and India combined in 2013. This is, in large part, due to the shutdown of all nuclear reactors in the country after the Fukushima Daiichi disaster (see the Japan Energy Policy discussion in section below).

| Fossil Fuel Balance of Trade (2013) (Import - Export) - Top 5 Polluters | | | |
|---|---|---|---|
| Country | Coal (Mt) | Oil (Mt) | Natural Gas (bm3) |
| China | 313 | 21 | 47 |
| United States | -99 | -81 | 37 |
| India | 191 | -53 | 17 |
| Russia | -114 | -115 | -204 |
| Japan | 197 | 29 | 123 |

*Color = Net Importer*
*All references are from 2013, just before the economic*
*turndown and indicative of the prior 8 years.*
*Enerdata "Global Energy Statistical Yearbook 2016"*

Importation of fossil fuels is conditional (especially with oil) on world market prices as well as overall consumption profiles.

There are many levels of analysis that can be performed using the simple data presented above. They are better handled within the discussion of the energy policy of the individual countries.

It would be a little more than disingenuous to present the four tables above without presenting one other for extremely serious consideration.

## Per Capita Energy Use
## (kg of oil equivalent)

| Country Name | 2013 | Rank |
|---|---|---|
| World | 1894.28 | |
| Qatar | 19120.34 | 1 |
| Iceland | 18177.25 | 2 |
| Trinidad and Tobago | 14537.57 | 3 |
| Curacao | 11800.98 | 4 |
| Bahrain | 10171.68 | 5 |
| Kuwait | 9757.45 | 6 |
| United Arab Emirates | 7691.01 | 7 |
| Brunei Darussalam | 7392.87 | 8 |
| Luxembourg | 7310.31 | 9 |
| Canada | 7202.23 | 10 |
| United States | 6915.84 | 11 |
| Norway | 6438.76 | 12 |
| Saudi Arabia | 6363.39 | 13 |
| Oman | 6232.46 | 14 |
| Finland | 6074.75 | 15 |
| Gibraltar | 5795.71 | 16 |
| Australia | 5586.34 | 17 |
| Korea, Rep. | 5253.47 | 18 |
| Sweden | 5131.54 | 19 |
| Russian Federation | 5093.06 | 20 |
| | ... | |
| Japan | 3570.44 | 32 |
| | ... | |
| China | 2226.27 | 53 |
| | ... | |
| India | 606.05 | 113 |

*Source: World Bank (last complete year re: consumption)*

*http://data.worldbank.org/indicator/EG.USE.PCAP.KG.OE*

*Retrieved 03/24/2017.*

Looking at per capita consumption is a leveling and sobering thought for anyone proposing solutions to the five largest consumers in the world (all in yellow highlighter). The United States is the worst offender on a per capita energy basis by a large margin.

# CHINA – ENERGY POLICY

At the end of the Chinese civil war Mao Zedong (Tse-tung) declared the existence of the People's Republic of China on October 1, 1949 from Tiananmen[77]. Since that time a consistent thread in energy policy in China has been a very strong correlation of economic growth and energy use. There may be an encouraging change in the rather unilateral strategy but it is very difficult to assess the exact reasons for the change and at which analytical level the official policy is derived.

There is a very interesting, although quite poorly written 12-page paper, produced by the Chinese Academy of Social Sciences (CASS) entitled "China Energy Outlook (2015-2016)".[78]  It presents scenarios in which there will be negative economic growth and an emphasis on energy efficiency gains, technology and alternative energy.  Within the context of a rigorously controlled output of government information and opinion this is a somewhat interesting and perhaps even encouraging document.  Its exact implications are difficult to gauge given the rather unsubstantial plans for the very ambitious changes.

There are two plausible and quite opposing possible analyses for these predications:

- China had already been feeling the global recession or at least a lessening of export potential both due to the recent global recession and increasing competition from developing countries.
- There is a fundamental understanding at higher levels in government that there are large negative environmental and social consequences of unbridled use of fossil fuels (especially coal) to fuel economic development.

The world would be very lucky if the driving force for economic downsizing was a rational global, humanistic and environmentally sensitive perspective (let us all have this fantastic dream!).

The more main-stream Chinese government sources are also using the same general rhetoric and approach as CASS in talking about the fundamental non-sustainability both economically and environmentally.  China's 13th 5-year-plan was approved at the fourth session of the 12th National People's Congress (NPC) that was held from March 5 to 16, 2016 in Beijing.  Among many other issues such as judicial reforms and supply-side structural economic reforms, the plans for energy featured technological innovation, efficiency gains, and increased investment in renewable energy[79].

---

[77] There is an approximately 4:21 minute video of the event on YouTube.
https://www.youtube.com/watch?v=TJcol3SJ6ww. **Retrieved 11/29/2016.**
[78] CASS Innovation Program *World Energy China Outlook 2015-2016, Interim Report.*
https://www.ief.org/_resources/files/snippets/chinese-academy-of-social-sciences-cass/world-energy-china-outlook-interim-report.pdf. **Retrieved 12/16/2016.**
[79] http://www.npc.gov.cn/npc/zgrdzz/site1/20160429/0021861abd66188d449902.pdf. **National People's Congress of China. "China's NPC Approves 13th Five-Year Plan". Issue 1, 2016. Retrieved 01/03/2017.**

"Facing downward economic pressure amid fragile global recovery, China's economy has entered what policymakers refer to as the "new normal," a phase of moderating growth driven more by consumption instead of exports and investment."[80]

Although the policy breaks some new ground it is somewhat vague on the operational means by which it will meet the goals.

Perhaps the severest critical comments on the 13th 5-year-plan come, not unexpectedly, from Greenpeace[81]. The list of criticisms was pointed and is abstracted below:

"The plan announced today fails to provide the much-needed blueprint for cleaning up China's power industry" … "Over the last three years China has seen coal consumption peak and record wind and solar installations. While the new plan consolidates many of these accomplishments, it does little to increase ambition."

"The plan limits coal-fired power generating capacity to 1,100 gigawatts in 2020, up from the current capacity of 920GW (*http://www.nea.gov.cn/xwfb/20161107zb1/index.htm*). Given that China already has severe overcapacity and demand for coal-fired power generation continues to fall, this is disappointing."

"Greenpeace is calling on the government to limit coal-fired capacity in 2020 at or below current level, by cancelling the vast glut of coal-fired power plants permitted in 2015, and accelerating retirements of existing coal plants. Investments in renewable energy should be further accelerated from 2015 levels to continue to reduce China's $CO_2$ emissions in line with what is required to combat climate change."

The criticisms leveled may indeed be legitimate but there are far more severe questions and criticisms that can be levelled and should be directed at the world's #2 polluter – the USA.

The present overcapacity for energy production in China very much parallels the problem in India where the economic downturn reduced demand but it is very difficult both physically and politically to halt projects that have taken years to develop and, by halting them, cause even more economic hardship. Both countries have had several years of high economic growth so the downturn is more apparent.

The real failure in the Chinese plan is the failure to aggressively use the downturn to realign future production by doing severe cuts in coal production while increasing solar and wind investments.

---

[80] Ibid, p. 18. Text written by Xinhua (China's official press agency) the biggest and most influential media organization in China.

[81] http://www.greenpeace.org/eastasia/press/releases/climate-energy/2016/China-Power-135YP/. "China's Power Sector 13th Five Year Plan disappoints – Greenpeace". Online publication 11/07/2016. Retrieved 01/03/2017.

# UNITED STATES – ENERGY POLICY

The central government voice for energy is the Department of Energy.[82] Depending upon ones' point of view, it may be fortunate or unfortunate that the United States has a legacy of freedom of expression that has promoted large numbers of conflicting viewpoints and quantities of contradictory information.

It seems totally hypocritical to complain about the wealth of information but there are secondary serious side effects of constant bombardment with information and opinion (often incorrect and misinformed). The average person has little or no real way to determine what is correct or not. This can lead to well-meaning but completely misguided support for poor policy or, perhaps worse, a complete withdrawal from the process due to the chaotic saturation. The second effect is all part of another relevant issue, "Bread and Circuses" (see a more comprehensive discussion in the section WHAT DOES THE POPE SAY? where there is a documentable leadership tactic of distraction that goes back to the 2nd century and likely further).

One of the primary motivational factors in US energy policy is the nature of the political system itself and the fact that every new president can strongly influence the management of every cabinet department by selecting a new person for the post of secretary. Within days of sitting down to re-write this section and after a couple of weeks collecting past data and trying to sift through it and prioritize it, there were two articles[83,84] warning of big changes at the Department of Energy under Donald Trump's appointee for Secretary – former Texas governor Rick Perry.

Rick Perry had previously stated that he wanted to get rid of the Department of Energy. It is difficult to understand his rationale, if there were indeed any, for that prior stance. As the author of this book I must confess that there are a myriad of problems at the Department of Energy but I would add to that the biggest problem is having a leader who is a staunch supporter of burning fossil fuel and a climate change naysayer. Apparently, the former Texas governor has understood, as many of the rest of us have, that US Department of Energy has not clearly articulated its multifunctional nature to the general public or at least not to Rick Perry.

Regardless of the specific viewpoints of the Secretary of Energy it will be important for the Energy Department to integrate that the leading producers cannot remain so indefinitely when their reserves

---

[82] http://www.energy.gov/.
[83] The New York Times. "Rick Perry, Ex-Governor of Texas, Is Trump's Pick as Energy Secretary" by CORAL DAVENPORT DEC. 13, 2016.
[84] National Geographic (blogs). "Changes Likely at Department of Energy under Trump Administration" Posted by Tim Profeta of Nicholas Institute for Environmental Policy Solutions, Duke University on December 15, 2016.

are insufficient. That will clearly be the case for the United States with its recent large increases in oil production due to fracking.[85] The numbers are not sustainable based on reserves as well as the operational limitation of increasing difficulty of production as fracking is used to retrieve products from more resistant geological formations.

This begs the question of whether there was any substantive energy policy involved at all in the substantial increase in oil production other than the environmental pass that was given to fracking to facilitate oil production.

Perhaps there is some merit in going back a little in time to March 30, 2011, when President Obama presented the 44-page "Blueprint for a Secure Energy Future"[86]. There is at least a temporal correlation to the substantial increases in US petroleum production. Below is the abstracted Table of Contents.

I.  **Introduction**

II.  **Executive Summary**

III.  **Develop and Secure America's Energy Supplies**

➤ Expand Safe and Responsible Domestic Oil and Gas Development and Production

➤ Lead the World Towards Safer, Cleaner, and More Secure Energy Supplies

IV.  **Provide Consumers with Choices to Reduce Costs and Save Energy**

➤ Reduce Consumer Costs at the Pump with More Efficient Cars and Trucks

➤ Cut Energy Bills with More Efficient Homes and Buildings

V.  **Innovate Our Way to a Clean Energy Future**

➤ Harness America's Clean Energy Potential

➤ Win the future through Clean Energy Research and Development

➤ Lead by Example: The Federal Government and Clean Energy

---

[85] Gallegos, T.J., and Varela, B.A., 2015, *Trends in hydraulic fracturing distributions and treatment fluids, additives, proppants, and water volumes applied to wells drilled in the United States from 1947 through 2010—Data analysis and comparison to the literature: U.S. Geological Survey Scientific Investigations Report 2014–5131*, 15 p., http://dx.doi.org/10.3133/sir20145131. ISSN 2328-0328 (online). First posted January 12, 2015 and Revised June 1, 2015. https://pubs.usgs.gov/sir/2014/5131/. Retrieved 02/09/2017.

[86] https://obamawhitehouse.archives.gov/sites/default/files/blueprint_secure_energy_future.pdf. Retrieved 02/09/2017. Due to the new Whitehouse administration these documents may change location.

I truly hate to put my stamp over the table of contents but the stamp is the only thing that one might believe. I am incredibly liberal (in fact, unashamedly leftist) and initially strongly supported the Obama presidency but my views were very much modified by the soft political pandering and somewhat misleading information in the document and in many other aspects of the Obama administration. This is one of the many reasons there is now a Trump presidency and a Republican party determined to undo at least Obamacare.

The "blueprint" reads fantastically well if you have no intention of examining any of the 44 pages of motherhood statements and many questionable scientific premises. This is very clearly a crafted political document and NOT a "Blueprint" since there are no real specifications, targets or measures. It often refers to unspecified advances in technology and innovation. It discusses safety without any specification of particular measures or real monitoring standards. The statement that there could really be an entity or substance called "Clean Coal" stacks up as one of the biggest pieces of political fecal matter generated prior to Trump's Whitehouse classifying any news that is not favorable "fake news".

As the new Secretary of Energy crafts what may be an intelligible energy policy, the world and global climate will just have to wait – and deteriorate even more measurably.

## RUSSIA – ENERGY POLICY

"The Energy policy of Russia is contained in an Energy Strategy document, which sets out policy for the period up to 2020. In 2000 the Russian government approved the main provisions of the Russian energy strategy to 2020, and in 2003 the new Russian energy strategy was confirmed by the government."

https://en.wikipedia.org/wiki/Energy_policy_of_Russia

The "factoid" above is not very helpful at all in trying to tease out a coherent Russian energy policy from the plethora of mixed political messages emanating primarily from Vladimir Putin.[87]  Regardless of his official title Putin has effectively been running Russia and shaping its policies since 1999.  Given his ability to retain power and wield it quite successfully, at least internally, it is critical to follow his international posturing to try to get a better handle on the general situation.

The bottom line is rather confusing.  Russia has a wealth of energy natural resources.  Much of the oil and gas is exported but the prices are often lower than OPEC prices.  This has certainly been a factor in the lower world oil prices that have prevailed until the most recent announcement of November 30, 2016 of an OPEC decision at a meeting in Vienna to reduce production by 300,000 barrels per day per member.  This is the first reduction in production in eight years.[88]  That caused a rapid rise in the market price in excess of $50/barrel.

It would be interesting to understand the reasoning behind the Russian approach.  One potential analysis is that Russia is using its resource base to place itself among the major world players in the energy game.  It may also translate into direct political influence if energy export deals are coupled to other agreements that may not be particularly overt.

The international oil market has become much more complicated in recent years.  A number of non-OPEC countries have become important players such as Canada (oil shale), the USA (fracking) and Russia.  In addition, a more stable situation on the ground in Iraq over the past year, and more importantly, the lifting of sanctions on Iran in January 2016[89] led to both countries pumping much more oil and that kept prices down as well.  It is reasonably clear that the prior dominance of Saudi Arabia in fixing the oil market by its dominance in OPEC has been substantially modified.

---

[87] Vladimir Putin first became Prime Minister in 1999, was Acting President 1999-2000, President 2000-2004, Second Presidency 2004-2008, Second Premiership (2008-2012), Third Presidency 2012-present.

[88] https://www.bloomberg.com/news/articles/2016-11-30/opec-said-to-agree-oil-production-cuts-as-saudis-soften-on-iran. Retrieved 12/08/2016.

[89] "Iran's Sanctions End as Deal Takes Effect", By CHRISTOPHER ALESSI and LAURENCE NORMAN, **The Wall Street Journal**, Updated Jan. 16, 2016 6:06 p.m. ET.  Retrieved 12/08/2016.

Since 2011 the USA has become the leading producer of fossil fuels in the world.  This is clearly shown in the chart below.  The production from Russia has been more-or-less consistent from about 2008 whereas US production has grown substantially.

**Estimated petroleum and natural gas hydrocarbon production in selected countries**
quadrillion British thermal units                                  million barrels per day of oil equivalent

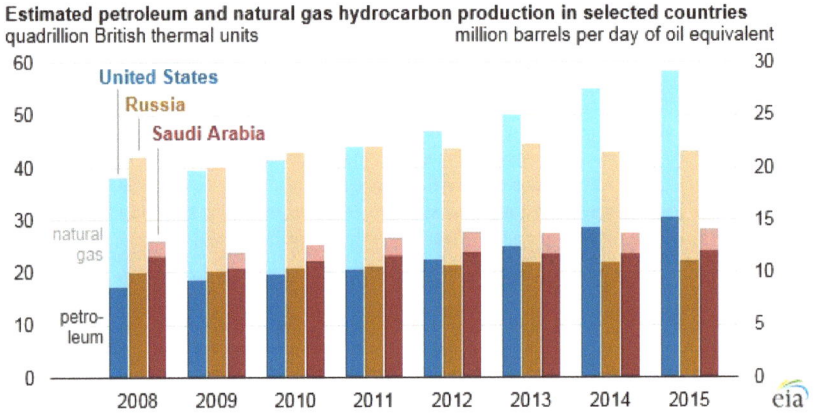

. Retrieved 12/08/2016.

If coal had been included in the chart above, the United States would be the leading country.  As in most cases, the devil is in the details.  The actual and predicted reserves as well as outstanding trade agreements and commitments are more important for future planning for any economy.

**NATURAL GAS RESERVES (proven)**

| Rank | Country | Natural gas proven reserves (m$^3$) | Date of information |
|---|---|---|---|
| 0 | World | 187,300,000,000,000 | |
| 1 | Russia | 48,700,000,000,000 | 12 June 2013 est.[5] |
| 2 | Iran | 33,600,000,000,000 | 12 June 2013 est.[7] |
| 3 | Qatar | 24,700,000,000,000 | June 2014.[8] |
| 4 | Turkmenistan | 17,500,000,000,000 | June 2014.[8] |
| 5 | United States | 9,860,000,000,000 | 12 December 2013[9][10] |
| 6 | Saudi Arabia | 8,600,000,000,000 | June 2014.[8] |
| 7 | Iraq | 6,400,000,000,000 | 1 January 2012 est.[11] |
| 8 | Venezuela | 5,724,500,000,000 | 19 July 2011[12] |
| 9 | Nigeria | 5,100,000,000,000 | June 2014.[8] |
| 10 | China | 4,643,000,000,000 | 1 January 2015[9] |
| 11 | Algeria | 4,502,000,000,000 | 1 January 2010 est. |
| 12 | Australia | 4,300,000,000,000 | 1 January 2014 est.[11] |
| 13 | Indonesia | 3,001,000,000,000 | 1 January 2010 est. |
| 14 | Azerbaijan | 2,600,000,000,000 | 1 January 2016 est.[13] |
| 15 | Malaysia | 2,350,000,000,000 | 1 January 2010 est. |

https://en.wikipedia.org/wiki/List_of_countries_by_natural_gas_proven_reserves
[REF.] are Wikipedia references.  Table truncated and retrieved 12/08/2016

**OIL RESERVES (proven)**

| Proven OIL Reserves (millions of barrels) | EIA (start of 2015) [1] | | OPEC (end of 2015) [2] | | BP (end of 2015) [3] | |
|---|---|---|---|---|---|---|
| | Rank | Reserves | Rank | Reserves | Rank | Reserves |
| Venezuela | 1 | 298,350 | 1 | 300,878 | 1 | 300,900 |
| Saudi Arabia | 2 | 268,289 | 2 | 266,455 | 2 | 266,600 |
| Canada | 3 | 172,481 | 26 | 4,118[4] | 3 | 172,200 |
| Iran | 4 | 157,800 | 3 | 158,400 | 4 | 157,800 |
| Iraq | 5 | 144,211 | 4 | 142,503 | 5 | 143,100 |
| Kuwait | 6 | 104,000 | 5 | 101,500 | 7 | 101,500 |
| UAE | 7 | 97,800 | 6 | 97,800 | 8 | 97,800 |
| Russia | 8 | 80,000 | 7 | 80,000 | 6 | 102,400 |
| Libya | 9 | 48,363 | 8 | 48,363 | 10 | 48,400 |
| United States | 10 | 39,933 | 10 | 36,685 | 9 | 55,000 |
| Nigeria | 11 | 37,070 | 9 | 37,062 | 11 | 37,100 |
| Kazakhstan | 12 | 30,000 | 11 | 30,000 | 12 | 30,000 |
| Qatar | 13 | 25,244 | 12 | 25,244 | 13 | 25,244 |
| China | 14 | 24,649 | 13 | 25,132 | 14 | 18,500 |
| Brazil | 15 | 15,314 | 14 | 16,184 | 15 | 13,000 |

Ref. [1]   U.S. EIA, International Energy Statistics, accessed 3 Sept. 2016.
Ref. [2]   OPEC, Annual Statistical Bulletin 2016
Ref. [3]   BP, Statistical Review of World Energy, June 2016.
Ref. [4]   OPEC does not include mined oil sand - under-reported Canada reserves

# OIL AND GAS INDUSTRY RESERVES BY NATION

**Key:** All oil numbers are in million barrels (MMbbl), gas numbers are in trillion cubic feet (Tcf)

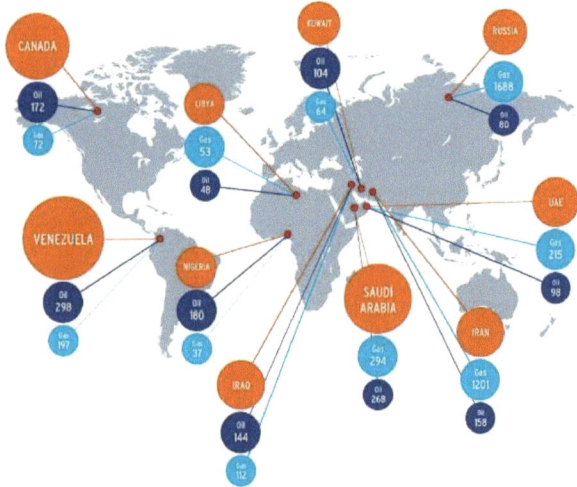

Statistic source  http://www.eia.gov/

## INDIA – ENERGY POLICY

As always, in researching any topic one must be very careful about sources and do routine cross-checks. India is a particularly good example when one considers the two almost diametrically opposed bits of data available (both retrieved from different Internet sources on 11/30/2016):

> "The energy policy of India is largely defined by the country's expanding energy deficit and increased focus on developing alternative sources of energy, particularly nuclear, solar and wind energy. The primary energy consumption in India is the third biggest after China and USA with 5.3% global share in 2015." https://en.wikipedia.org/wiki/Energy_policy_of_India

AND

> "NEW DELHI: India won't need any new power plants for the next three years as it is flush with generation capacity, according to a government assessment." As reported by ET Bureau | Updated: Jun 02, 2016, 05.47 AM IST.
>
> http://economictimes.indiatimes.com/industry/energy/power/india-wont-need-extra-power-plants-for-next-three-years-says-government-report/articleshow/52545715.cms

The stamp is completely warranted given the simple fact that one can't simultaneously have an energy deficit and an energy surplus – it is necessary to do the extra work or thinking to determine why. The Wikipedia entry for "Energy Policy of India" starts with an assertion that the country has an "expanding energy deficit" but the article in the Economic Times of India contradicts this completely. Why? Well the answer is as simple as timing. The last few years have seen a global economic slowdown and what seemed like a deficit in power will become a surplus if demand scales down in an unpredictable level and manner. The Wikipedia entry is dated and, by necessity not representative of the actual situation.

For many years India had been riding, and sometimes leading the global economic surge but despite the recent global slowdown it seems to be doing very well[90]. Infrastructure projects like power plants require considerable startup time and cannot be "turned off" by simple acts of political will or chicanery. Therefore, it should be absolutely no surprise that there is apparently excess power capacity in a period of economic downturn.

The essential elements of Indian energy policy are still very simple. There is a direct correlation of power consumption and economic growth. This seems to be one of the intrinsic hazards of living within a democratic political system where elected representatives have a very large influence on all resource

---

[90] "IMF revises India's GDP forecast to 7.6% on robust growth momentum" by Asit Ranjan Mishra (first published October 4, 2016. "India's economy grew 7.6% in 2015-16 and the Narendra Modi government expects it to grow close to 8% in 2016-17. http://www.livemint.com/Politics/UIDIyxwzncrZFdBiChFrwL/IMF-revises-Indias-GDP-forecast-to-76-on-robust-growth-mo.html. Retrieved 01/10/2017.

and development decisions.  Any policy that looks like it is not numerically "growing" the economy will almost invariably become political suicide.  Alas, there is but one Gandhi per generation, maybe not even that many and that is sad for all of us – vision is costly.

## JAPAN – ENERGY POLICY

Japanese energy policy can best (and most kindly) be typified as reactionary and overcompensating. This is a very direct effect of the Fukushima Daiichi nuclear disaster (March 11, 2011). In the section on nuclear accidents there is some time spent on discussing why this was much less a nuclear disaster than a situation of incredibly poor management and economic / business concerns trumping technical /scientific ones. There are, most surely, meta-issues involved as well. Most important is the incredible sensitivity of the Japanese people to the fact that they have been the only ones who have been the recipients of nuclear explosions. A visit to the Hiroshima Peace Memorial Museum[91] will help to frame the cultural zeitgeist (the "spirit" or "ghost" of the time).

The Japanese government's response to Fukishima must be considered extreme. An energy white paper, approved by the Japanese Cabinet in October 2011, stated that "public confidence in safety of nuclear power was greatly damaged" by the Fukushima disaster, and called for a reduction in the nation's reliance on nuclear power.[92]

This resulted in the effective shutdown of fifty plus reactors for safety inspections. This was completely reactionary and political in nature. The very politicians who had not properly regulated their own nuclear industry then shut it down based on a lack of public confidence that they had materially contributed strongly to by not calling responsible managers of facilities into question and by caving in completely to "public opinion" instead of modifying and shaping it. It is somewhat surprising that someone did not impale themselves on a katana – perhaps a viable option for the manager(s) of Tepco at Fukushima Daiichi.

The table on the next page has the last column highlighted in yellow and the header is "UCF for 2014" and all the numbers in the column are 0. UCF is the short form for:

> "Unit capability factor is defined as the ratio of the available energy generation over a given time period to the reference energy generation over the same time period, expressed as a percentage."[93]

---

[91] https://en.wikipedia.org/wiki/Hiroshima_Peace_Memorial_Museum. Built in 1955 and remodeled in 1994. Retrieved 12/01/2016.

[92] Tsuyoshi Inajima & Yuji Okada (Oct 28, 2011). "Nuclear Promotion Dropped in Japan Energy Policy After Fukushima". *Bloomberg*.

[93] https://www.iaea.org/PRIS/Glossary.aspx. There is a more expansive definition but this will suffice. Retrieved 12/08/2016.

# IAEA Country Nuclear Profiles 2015 - Table for Japan

## TABLE 7. STATUS AND PERFORMANCE OF NUCLEAR POWER PLANTS

| Reactor Unit | Type | Net Capacity [MW(e)] | Status | Operator | Reactor Supplier | Construction Date | First Criticality Date | First Grid Date | Commercial Date | Shutdown Date | UCF for 2014 |
|---|---|---|---|---|---|---|---|---|---|---|---|
| FUKUSHIMA-DAINI-1 | BWR | 1067 | Operational | TEPCO | TOSHIBA | 03/16/76 | 06/17/81 | 07/31/81 | 04/20/82 | | 0 |
| FUKUSHIMA-DAINI-2 | BWR | 1067 | Operational | TEPCO | HITACHI | 05/25/79 | 04/26/83 | 06/23/83 | 02/03/84 | | 0 |
| FUKUSHIMA-DAINI-3 | BWR | 1067 | Operational | TEPCO | TOSHIBA | 03/23/81 | 10/18/84 | 12/14/84 | 06/21/85 | | 0 |
| FUKUSHIMA-DAINI-4 | BWR | 1067 | Operational | TEPCO | HITACHI | 05/28/81 | 10/24/86 | 12/17/86 | 08/25/87 | | 0 |
| GENKAI-1 | PWR | 529 | Operational | KYUSHU | MHI | 09/15/71 | 01/28/75 | 02/14/75 | 10/15/75 | | 0 |
| GENKAI-2 | PWR | 529 | Operational | KYUSHU | MHI | 02/01/77 | 05/21/80 | 06/03/80 | 03/30/81 | | 0 |
| GENKAI-3 | PWR | 1127 | Operational | KYUSHU | MHI | 06/01/88 | 05/28/93 | 06/15/93 | 03/18/94 | | 0 |
| GENKAI-4 | PWR | 1127 | Operational | KYUSHU | MHI | 07/15/92 | 10/23/96 | 11/12/96 | 07/25/97 | | 0 |
| HAMAOKA-3 | BWR | 1056 | Operational | CHUBU | TOSHIBA | 04/18/83 | 11/21/86 | 01/20/87 | 08/28/87 | | 0 |
| HAMAOKA-4 | BWR | 1092 | Operational | CHUBU | TOSHIBA | 10/13/89 | 12/02/92 | 01/27/93 | 09/03/93 | | 0 |
| HAMAOKA-5 | BWR | 1325 | Operational | CHUBU | TOSHIBA | 07/12/00 | 03/23/04 | 04/30/04 | 01/18/05 | | 0 |
| HIGASHI DORI-1 (TOHOKU) | BWR | 1067 | Operational | TOHOKU | TOSHIBA | 11/07/00 | 01/24/05 | 03/09/05 | 12/08/05 | | 0 |
| IKATA-1 | PWR | 538 | Operational | SHIKOKU | MHI | 09/01/73 | 01/29/77 | 02/17/77 | 09/30/77 | | 0 |
| IKATA-2 | PWR | 538 | Operational | SHIKOKU | MHI | 08/01/78 | 07/31/81 | 08/19/81 | 03/19/82 | | 0 |
| IKATA-3 | PWR | 846 | Operational | SHIKOKU | MHI | 10/01/90 | 02/23/94 | 03/29/94 | 12/15/94 | | 0 |
| KASHIWAZAKI KARIWA-1 | BWR | 1067 | Operational | TEPCO | TOSHIBA | 06/05/80 | 12/12/84 | 02/13/85 | 09/18/85 | | 0 |
| KASHIWAZAKI KARIWA-2 | BWR | 1067 | Operational | TEPCO | TOSHIBA | 11/18/85 | 11/30/89 | 12/08/90 | 09/28/90 | | 0 |
| KASHIWAZAKI KARIWA-3 | BWR | 1067 | Operational | TEPCO | TOSHIBA | 03/07/89 | 10/19/92 | 12/08/92 | 08/11/93 | | 0 |
| KASHIWAZAKI KARIWA-4 | BWR | 1067 | Operational | TEPCO | HITACHI | 03/05/90 | 11/01/93 | 12/21/93 | 08/11/94 | | 0 |
| KASHIWAZAKI KARIWA-5 | BWR | 1067 | Operational | TEPCO | HITACHI | 06/20/85 | 07/20/89 | 09/12/89 | 04/10/90 | | 0 |
| KASHIWAZAKI KARIWA-6 | BWR | 1315 | Operational | TEPCO | TOSHIBA | 11/03/92 | 12/18/95 | 01/29/96 | 11/07/96 | | 0 |
| KASHIWAZAKI KARIWA-7 | BWR | 1315 | Operational | TEPCO | HITACHI | 07/01/93 | 11/01/96 | 12/17/96 | 07/02/97 | | 0 |
| MIHAMA-1 | PWR | 320 | Operational | KEPCO | WH | 02/01/67 | 07/29/70 | 08/08/70 | 11/28/70 | | 0 |
| MIHAMA-2 | PWR | 470 | Operational | KEPCO | MHI | 05/29/68 | 04/10/72 | 04/21/72 | 07/25/72 | | 0 |
| MIHAMA-3 | PWR | 780 | Operational | KEPCO | MHI | 08/07/72 | 01/28/76 | 02/19/76 | 12/01/76 | | 0 |
| OHI-1 | PWR | 1120 | Operational | KEPCO | WH | 10/26/72 | 12/02/77 | 12/23/77 | 03/27/79 | | 0 |
| OHI-2 | PWR | 1120 | Operational | KEPCO | WH | 12/08/72 | 09/14/78 | 10/11/78 | 12/05/79 | | 0 |
| OHI-3 | PWR | 1127 | Operational | KEPCO | MHI | 10/03/87 | 05/17/91 | 06/07/91 | 12/18/91 | | 0 |
| OHI-4 | PWR | 1127 | Operational | KEPCO | MHI | 06/13/88 | 05/28/92 | 06/19/92 | 02/02/93 | | 0 |
| ONAGAWA-1 | BWR | 498 | Operational | TOHOKU | TOSHIBA | 07/08/80 | 10/18/83 | 11/18/83 | 06/01/84 | | 0 |
| ONAGAWA-2 | BWR | 796 | Operational | TOHOKU | TOSHIBA | 04/12/91 | 11/02/94 | 12/23/94 | 07/28/95 | | 0 |
| ONAGAWA-3 | BWR | 796 | Operational | TOHOKU | TOSHIBA | 01/23/98 | 04/26/01 | 05/30/01 | 01/30/02 | | 0 |
| SENDAI-1 | PWR | 846 | Operational | KYUSHU | MHI | 12/15/79 | 08/25/83 | 09/16/83 | 07/04/84 | | 0 |
| SENDAI-2 | PWR | 846 | Operational | KYUSHU | MHI | 10/12/81 | 03/18/85 | 04/05/85 | 11/28/85 | | 0 |
| SHIKA-1 | BWR | 505 | Operational | HOKURIKU | HITACHI | 07/01/89 | 11/20/92 | 01/12/93 | 07/30/93 | | 0 |
| SHIKA-2 | BWR | 1108 | Operational | HOKURIKU | HITACHI | 08/20/01 | 05/26/05 | 07/04/05 | 03/15/06 | | 0 |
| SHIMANE-1 | BWR | 439 | Operational | CHUGOKU | HITACHI | 07/02/70 | 06/01/73 | 12/02/73 | 03/29/74 | | 0 |
| SHIMANE-2 | BWR | 789 | Operational | CHUGOKU | HITACHI | 02/02/85 | 05/25/88 | 07/11/88 | 02/10/89 | | 0 |
| TAKAHAMA-1 | PWR | 780 | Operational | KEPCO | WH/MHI | 04/25/70 | 03/14/74 | 03/27/74 | 11/14/74 | | 0 |
| TAKAHAMA-2 | PWR | 780 | Operational | KEPCO | MHI | 03/09/71 | 12/20/74 | 01/17/75 | 11/14/75 | | 0 |
| TAKAHAMA-3 | PWR | 830 | Operational | KEPCO | MHI | 12/12/80 | 04/17/84 | 05/09/84 | 01/17/85 | | 0 |
| TAKAHAMA-4 | PWR | 830 | Operational | KEPCO | MHI | 03/19/81 | 10/11/84 | 11/01/84 | 06/05/85 | | 0 |
| TOKAI-2 | BWR | 1060 | Operational | JAPCO | GE | 10/03/73 | 01/18/78 | 03/13/78 | 11/28/78 | | 0 |
| TOMARI-1 | PWR | 550 | Operational | HEPCO | MHI | 04/18/85 | 11/16/88 | 12/06/88 | 06/22/89 | | 0 |
| TOMARI-2 | PWR | 550 | Operational | HEPCO | MHI | 06/13/85 | 07/25/90 | 08/27/90 | 04/12/91 | | 0 |
| TOMARI-3 | PWR | 866 | Operational | HEPCO | MHI | 11/18/04 | 03/03/09 | 03/20/09 | 12/22/09 | | 0 |
| TSURUGA-1 | BWR | 340 | Operational | JAPCO | GE | 11/24/66 | 10/03/69 | 11/16/69 | 03/14/70 | | 0 |
| TSURUGA-2 | PWR | 1108 | Operational | JAPCO | MHI | 11/06/82 | 05/28/86 | 06/19/86 | 02/17/87 | | 0 |
| OHMA | BWR | 1325 | Under Construction | EPDC | H/G | 05/07/10 | | | | | |
| SHIMANE-3 | BWR | 1325 | Under Construction | CHUGOKU | HITACHI | 10/12/07 | | | | | |
| MONJU | FBR | 246 | Operational | JAEA | T/H/F/M | 05/10/86 | 04/05/94 | 08/29/95 | | | 0 |
| FUGEN ATR | HWLWR | 148 | Permanent Shutdown | JAEA | HITACHI | 05/10/72 | 03/20/78 | 07/29/78 | 03/20/79 | 03/29/03 | |
| FUKUSHIMA-DAIICHI-1 | BWR | 439 | Permanent Shutdown | TEPCO | GE/GETSC | 07/25/67 | 10/10/70 | 11/17/70 | 03/26/71 | 05/19/11 | |
| FUKUSHIMA-DAIICHI-2 | BWR | 760 | Permanent Shutdown | TEPCO | GE/T | 06/09/69 | 05/10/73 | 12/24/73 | 07/18/74 | 05/19/11 | |
| FUKUSHIMA-DAIICHI-3 | BWR | 760 | Permanent Shutdown | TEPCO | TOSHIBA | 12/28/70 | 09/06/74 | 10/26/74 | 03/27/76 | 05/19/11 | |
| FUKUSHIMA-DAIICHI-4 | BWR | 760 | Permanent Shutdown | TEPCO | HITACHI | 02/12/73 | 01/28/78 | 02/24/78 | 10/12/78 | 05/19/11 | |
| FUKUSHIMA-DAIICHI-5 | BWR | 760 | Permanent Shutdown | TEPCO | TOSHIBA | 05/22/72 | 08/26/77 | 09/22/77 | 04/18/78 | 12/17/13 | |
| FUKUSHIMA-DAIICHI-6 | BWR | 1067 | Permanent Shutdown | TEPCO | GE/T | 10/26/73 | 03/09/79 | 05/04/79 | 10/24/79 | 12/17/13 | |
| HAMAOKA-1 | BWR | 515 | Permanent Shutdown | CHUBU | TOSHIBA | 06/10/71 | 06/20/74 | 08/13/74 | 03/17/76 | 01/30/09 | |
| HAMAOKA-2 | BWR | 806 | Permanent Shutdown | CHUBU | TOSHIBA | 06/14/74 | 03/28/78 | 05/04/78 | 11/29/78 | 01/30/09 | |
| JPDR | BWR | 12 | Permanent Shutdown | JAEA | GE | 12/01/60 | 08/22/63 | 10/26/63 | 03/15/65 | 03/18/76 | |
| TOKAI-1 | GCR | 137 | Permanent Shutdown | JAPCO | GEC | 03/01/61 | 05/04/65 | 11/10/65 | 07/25/66 | 03/31/98 | |

Data source: IAEA - Power Reactor Information System (PRIS).

Note: Table 7 is completely generated from PRIS data to reflect the latest available information and may be more up to date than the text of the report.
http://www-pub.iaea.org/MTCD/Publications/PDF/CNPP2015_CD/countryprofiles/Japan/Japan.htm

12/5/2016

What the numbers mean is that **in 2014 ALL of the 54 "OPERATIONAL" reactors in Japan were actually not operating** ... i.e. doing their designed jobs of producing power. The UFCs should be in the 90's. The lost capacity is 42634 MW that Japan has chosen to make up by burning fossil fuel!

Closing all the reactors in any country is a serious decision that should be made on purely technical grounds not on gut reaction, irrational fear, and political pandering. **There is no technical or fundamental argument that could possibly back such an extreme reaction**. The statistics alone of 54 reactors suffering from serious safety problems meriting shutdown and inspection would be incredibly low even in the most lax regulatory environment – certainly not Japan.

In general, one can confidently state that the nuclear power industry is one of the most highly regulated and safest in the world! On the surface this may seem like an outrageous overstatement but I believe it can be said with utter seriousness. The grave and "irrational" fear of "nuclear events" virtually guarantees much closer scrutiny than traditional forms of providing energy such as coal which, in-and-of-itself is a huge killer of people both directly and indirectly.

If it is not already clear I personally think that the Japanese reaction to Fukushima is incredibly mutated – and in a bad way – like Godzilla or Mothra. Instead of weeding out poor executive and management decisions which led to a completely predictable and avoidable event, the government decide to gut an industry that the country and the world at large desperately needs to function efficiently. ***This is not energy policy it is energy folly!***

Not only can Japan not live up to its commitments to the Kyoto accords, the energy path chosen will elongate the economic woes of Japan for many years in addition to increasing its carbon footprint and already large contribution to global warming. Of the 5 leading nations in consumption Japan has the lowest natural resource base and the greatest need for nuclear power!

## PRINCIPLES FOR AN IDEAL ENERGY POLICY

What would constitute an ideal energy policy? This is not meant as a trivial, rhetorical challenge or a monumental task but a thought experiment if we are permit ourselves to do so!

Energy production and consumption constitute a global phenomenon and there are many consequences of this relatively simple observation.

- Energy policy should be directed at a global rather than national level
  - o Acceptance of this principle would require co-development and sharing of resources.
- A primary concern should be to provide energy with as little environmental impact as possible.
  - o Large-scale fossil fuel burning must come to an end as soon as possible!
    - ▪ If this occurs the buffering capacity of the planet could then handle reasonable biomass conversion to energy.
  - o All energy calculations and presentations should be normalized on a *per capita* basis.
  - o The largest *per capita* energy consumers should carry the largest proportionate economic burden to reduce global $CO_2$.
- A secondary concern is the energy consumed in product creation and areas of enormous energy wastage such as in consumer goods packaging.
- All nations must show positive reductions in $CO_2$ proportionate to *per capita* consumption.
- All nations must show positive progress towards clean / renewable energy that does not increase $CO_2$ production.[94]

---

[94] It is critical to note that biofuels such as corn-based ethanol do contribute to $CO_2$ as does the biomass burning of organic waste **and these must be scaled accordingly**.

# THE REAL COSTS OF POWER PRODUCTION

*"The price of anything is the amount of life you exchange for it."*

Henry David Thoreau, American Author (July 12, 1817 – May 6, 1962)

How much life have we already exchanged and are we willing to exchange to produce and use power?

*"Cecil Graham: What is a cynic?*
*Lord Darlington: A man who knows the price of everything, and the value of nothing."*

Oscar Wilde, *Lady Windermere's Fan*, Irish Playwright (October 16, 1854 - November 30, 1900)

The meta-irony of using these quotations to introduce this subject is that very few people have ever considered the "real" costs of anything! It's important to articulate what is meant by this. Perhaps the essential point is that we rarely, if ever, have a clear and objective view of the elements and processes in coming to any decision or, for that matter, in creating any object or process in the real world.

All industries and endeavors exhibit enlightened self-interest in representing themselves in the most positive light. This is intrinsically understandable but an almost inevitable consequence is that an objective and balanced picture is often not available or is even actively obscured.

The production and consumption of power should (... MUST) contain all of the elements involved in finding, securing, delivering, and mitigating the effects of energy use.

**Extraction** – Actual energy used to produce the raw "fuel". Environment issues ... construction, pollution, habitat disruption and loss. Health issues ... deaths.

**Refining / purifying** - Actual energy used to produce the usable "fuel". Environmental issues ... effluent streams, waste treatment and disposal. Health issues.

**Consumption (burning)** – Environment issues ... $CO_2$ burden, other gaseous pollutants, atmospheric aerosols, solid waste products such as fly-ash). Climate change ... agricultural consequences, marine consequences (ocean acidification).

One stunning example of a problem of presentation and perception is given below. I refuse, in advance, to apologize for anything that I may satirize. I take it as a personal affront to my intelligence to read the legalese and political double-speak (mostly from governments) that is exceptionally prevalent when discussing energy and the environment.

## ELECTRIC CARS – THE GREAT "SHELL GAME"

I recently had a discussion (really, it was more of my ranting) about electric cars with an affable, young, environmental studies graduate, working for the California Air Quality Management District (AQMD).

Tesla – those beautiful electric cars all come out of the dealership with the proudly displayed license plate placards "Zero Emissions". **How, in any universe, is this possible?** This is a parsing or willful misuse of language rather than an understanding of energy and energy efficiency.

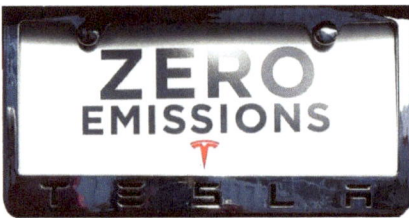

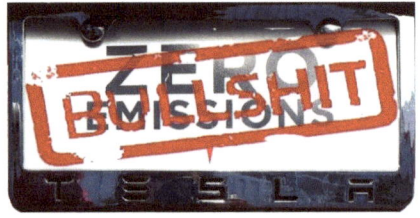

Where does the power come from that runs the car? Many people act like Homer Simpson and say, "DOH! Are you dumb or something? The wall plug, of course!" Yes, we should all be acutely aware of wall plugs and of how "clean" electricity is – *about as clean as the fossil fuels used to create the majority of it!*

The following data was taken directly from the EIA website 02/15/2017: (*https://www.eia.gov/tools/faqs/faq.cfm?id=427&t=3*). They list the last update as 04/01/2016.

Major energy sources and percent share of total U.S. electricity generation in 2015:

- Coal                    = 33%
- Natural gas             = 33%
- Nuclear                 = 20%
- Hydropower              = 6%
- Other renewables        = 7% (Biomass = 1.6%, Geothermal = 0.4%, Solar = 0.6%, Wind = 4.7%)
- Petroleum               = 1%
- Other gases             = <1%

Preliminary data; based on generation by utility-scale facilities.

About 67% of all electric power in the United States is from fossil fuels. So … fossil fuels are burnt in a power plant generally many miles away from where the electricity is transported. There are some fundamental issues that must be addressed and the most important one is line loss of electricity in transmission.

Line loss of power is fundamentally due to an effect encapsulated in Joule's Law that states that energy losses are directly proportional to the square of the current. Reducing the current by a factor of two will lower the energy lost to resistance by a factor of four for any conductor. So how is current reduced? By upping the voltage (Power(P) = Voltage(V) * Current(I)). That is why any long-distance power transmission lines are generally high voltage. "At the power stations, the power is produced at a relatively low voltage between about 2.3 kV and 30 kV, depending on the size of the unit. The generator terminal voltage is then stepped up by the power station transformer to a higher voltage (115 kV to 765 kV AC, varying by the transmission system and by country) for transmission over long distances."[95]

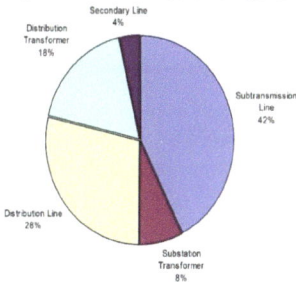

Component Contribution to 5.3% Overall Energy Loss

The power is produced at low voltage, stepped up to high voltage for long-distance transmission, stepped down (transformed) to low voltage for local distribution / consumption. The line losses in major transmissions lines are relatively low but increase substantially when the voltage is stepped down for local distribution. The following diagram and chart were taken from a 2007 report done in Canada for Hydro One in Ontario.[96]

This report was done for Hydro One (effectively, formerly Ontario Hydro which is the government owned electric power utility). This particular source was used based on prior experience with the general veracity of research materials that came out

| Customer Type | TLF in Present Rates | New Estimate of Technical Losses (2007 study) | New Estimate of DLF (2007 study) | New Estimate of TLF (2007 study) |
|---|---|---|---|---|
| Embedded LDC and Subtransmission Customers | 3.4% | 2.6% | 3.8% | 4.4% |
| Primary Customers | 6.1% | 5.6% | 6.8% | 7.4% |
| Secondary Customers | 9.1% | 8.2% | 9.4% | 10.0% |

\* Note: The TLFs include technical losses and non-technical losses on the distribution system and the supply facilities loss factor (0.6%) for losses on the transmission system supply transformer. In the Present Rates in column 2 the non-technical losses are estimated as 10% of the technical losses. In the new 2007 analysis in columns 4 and 5, non-technical losses are included as 1.2% of the energy sold.

[95] The Wikipedia entry on power transmission and distribution is quite well done and properly informative. https://en.wikipedia.org/wiki/Electric_power_transmission.
[96] 2007 RECALCULATION OF DISTRIBUTION SYSTEM ENERGY LOSSES AT HYDRO ONE. Kinectrics Inc. Report No: K-013111-001-RA-0001-R01. July 27, 2007. Ray Piercy, Senior Engineer, Transmission and Distribution Technologies: Stephen L. Cress, Manager – Distribution Systems, Transmission and Distribution Technologies. Page viii in Report K-013111-001-RA-0001-R01. Retrieved as .pdf 11/17/2016 from http://www.hydroone.com/RegulatoryAffairs/Documents/EB-2007-0681/Exhibit%20A/Tab_15_Schedule_3_Distribution_Line_Losses_Study.pdf.

or were solicited by Ontario Hydro ca. 1970's – 1980's).

The data clearly shows that in one of the most efficient power systems in the world (Canada) **the fundamental line losses of electricity to primary customers are 7.4% and 10.0% to secondary customers.**

The real losses are considerably higher for normal domestic users in the distributed system (some reports are up to 20%). It is also highly likely that there is even additional loss at the charger-car battery interface.

So now the question must be reformulated. Given that electric power losses probably at least 10% then is it appropriate to decrease the efficiency of any original source appropriately? I personally and scientifically believe that the answer is YES.

Given the above conclusion if the source of the electricity is primarily from burning fossil fuel then the emissions of the electric car are equivalent to the burnt fossil fuel emissions at origin plus (a minimum) of 10%. **MEANING ELECTRIC CARS ARE DIRTIER THAN EQUIVENT POWERED CARS USING FOSSIL FUEL DIRECTLY! (as long as the major source of electricity is dirty)**. This is a simple operational conclusion. Perhaps someone at the Department of Energy has found a Nobel Prize winning new technology that would negate what logical people would consider the facts. Perhaps they have joined the Trump administration with an entire set of "Alternative Facts".

There are several possible descriptors for anyone who has signed on to this approach for selling "electric cars" but, as a wise editor told me, I should not digress completely towards negative ranting in describing such people. It should be sufficient to say that they have not looked at the facts and/or integrated them. They are acting as if they were indeed Homer Simpson.

Electric cars will only be clean when they are primarily (preferably completely) powered by clean power - not fossil fuels. There are extremely few countries in the world where the claim that electric cars were relatively clean would be credible – Sweden, Denmark, Scotland, and Iceland are immediate candidates. This is all very admirable but, again, statistically insignificant.

Sweden has set a very ambitious goal to be 100% fossil fuel free[97] and may well be the first country to achieve this. This effort will take until at least 2030 if one "reads between the lines" but it is a noble

---

[97] http://www.government.se/articles/2016/07/a-climate-policy-framework-and-a-climate-and-clean-air-strategy-

effort. So, where did the notion that electric cars were "Zero Emissions" come from? The energy fairy? Perhaps this is not a bad guess because it is likely the US Department of Energy itself.

When one looks carefully (as a lawyer would) at what is written on DOE websites, in particular the Alternative Fuels Data Center[98] one can read about:

> **"Emissions**
>
> Hybrid and plug-in electric vehicles can have significant emissions benefits over conventional vehicles. HEV emissions benefits vary by vehicle model and type of hybrid power system. EVs produce zero tailpipe emissions, and PHEVs produce no tailpipe emissions when in all-electric mode.
>
> The life cycle emissions of an EV or PHEV depend on the sources of electricity used to charge it, which vary by region. In geographic areas that use relatively low-polluting energy sources for electricity production, plug-in vehicles typically have a life cycle emissions advantage over similar conventional vehicles running on gasoline or diesel. In regions that depend heavily on conventional fossil fuels for electricity generation, PHEVs and EVs may not demonstrate a strong life cycle emissions benefit. Use the Vehicle Cost Calculator to compare life cycle emissions of individual vehicle models in a given location."

The key word here is "**tailpipe**". It is true that electric vehicles have "Zero **Tailpipe** Emissions" *NOT* "Zero Emissions" when one considers where the energy is actually coming from … i.e. how it is produced. The second part has the author's stamp because of the evasive lawyerly manner in which it was worded. If it was indeed worded by a scientist, I would personally vote to have that scientist publicly executed for extreme intellectual dishonesty or, at minimum, flogged publicly, for stretching truth and credibility to the breaking point. Presently, there is no state in the United States where the source of the electricity could ever make the car clean!

This electric vehicle emissions game is another manifestation of NIMBY syndrome (**N**ot **I**n **M**y **B**ack **Y**ard). As long as emissions are not coming out of the tail pipe but in a fossil fuel burning power plant, perhaps in the Ohio River valley, it is perfectly fine to funnel huge amounts of $CO_2$ into the atmosphere and send them downwind.

In fact, the government is heavily subsidizing this sham and the real problem is that many people are buying into it. Instead of making clear that there needs to be real and proportionate investment in "future clean technologies" that are **not now efficient** (but will be if we pursue the "correct" course) and appealing to the need for constraints and measures to stop burning fossil fuels and biomass (any $CO_2$ emitters) the powers that be have decided to ignore all measures of objective reality.

---

for-sweden/. Retrieved 11/29/2016.
[98] Department of Energy website http://www.afdc.energy.gov/. Passage retrieved 11/21/2016.

The Department of Energy (see US Energy policy for more substantive comment) is not accurately or properly representing many issues from electric cars, per capita energy use, through "clean coal" (an oxymoron, at best, since there is no such thing as clean coal). The EIA postings and information are often more politically motivated than scientifically appropriate.

There may well be a large component of creative deception in the actions of the US Department of Energy. For example [author projection of a DOE staff meeting] … "Why don't we see how far into the massive amount of useless circular verbiage we have created before people actually give up trying to make sense of it?" When explanations are not clear, concise and referenced, then one must ask what the real intent of those messages is.

To add to the levels of misinformation, there are countless city, county and other buses that brag, in large writing on their exterior, that they are "clean air vehicles" when they are using Compressed Natural Gas (CNG). So, did someone in government or, more likely, in the PR firm find another new, Nobel Prize level way to make CNG into something other than a fossil fuel?? That would be astounding and then there would be no reason to be worried about the climate crisis.

## WHAT DOES THE POPE SAY?

> *"We know that technology based on the use of highly polluting fossil fuels –*
> *especially coal, but also oil and, to a lesser degree, gas – needs to*
> *be progressively replaced without delay. Until greater progress is made in*
> *developing widely accessible sources of renewable energy, it is legitimate*
> *to choose the lesser of two evils or to find short-term solutions. But the*
> *international community has still not reached adequate agreements about*
> *the responsibility for paying the costs of this energy transition."* Laudato si' 165. P. 59.[99]

Pope Francis (Jorge Mario Bergoglio, December 17, 1936 – present.  He is the 266th and current Pope of the Catholic Church and Sovereign of the Vatican City)

While struggling with the tenor and direction of this book I was blindsided by my own ignorance and found myself reading and being completely impressed with Pope Francis's "*Laudato si'*.  ENCYCLICAL LETTER OF THE HOLY FATHER ON CARE FOR OUR COMMON HOME" [the Earth] May 24, 2015.  In my former life as a scientist I would not have ever have been caught reading material so nominally away from mainstream science.  However, my own learning has been tempered by the real world and the need to gather information from as many sources as possible.  The document was masterfully credible both from a scientific and a political / human perspective.  It is the later aspect that is the most impressive in that **a political and social analysis of the climate crisis is more instructive and potentially helpful than a purely scientific approach.**

In the first quotation (above) there is a remarkably practical, simple, and accurate definition of the present situation along with one of many of the associated problems with the evolution of any potential solution - *"the international community has still not reached adequate agreements about the responsibility for paying the costs of this energy transition."*

Given the five major polluters (China, United States, India, Russia, Japan) and the critical importance of their operational *modus operandi* it seemed not only logical but imperative to attempt to understand any possible causative links (political, ideological, historic, economic, and practical / operational) between them.  The political differences in the countries run to the extreme – from capitalist to what is still primarily communist.

Looking first at political (and implicitly, ideological) comparisons the five countries can be relatively simplistically summarized below.

---

[99] http://w2.vatican.va/content/francesco/en/encyclicals/documents/papa-francesco_20150524_enciclica-laudato-si.htm.  Retrieved 06/25/2015.

Regardless of its communist underpinnings China has become fundamentally driven by economic forces and concerns that are not generally aligned with communism. It may be a moot point but it is necessary to attempt to be accurate in terms of typifying the present Chinese political / economic system. In an economically pure sense the Chinese economic system, as it presents itself in 2016-17, is truly more mercantile than capitalist[100]. The visible growing middle class and the spike in consumerism as well as in production of materials desired in the west have all fueled the interest in foreign trade, balance of payments, currency manipulation, and all the familiar trappings of any capitalist power. However, the means of production and the control over it is still very much under strict political direction.

Although the United States is often characterized as an example of free market capitalism it is more accurate to typify the US as a "mixed economy". There is a high degree of diversified private ownership and individual freedom in determining prices and services but the government accounts for a significant portion of the economy (approximately one-third). Prior to the Great Depression of the 1930's the government had a minimal economic stake and the system was indeed primarily free-market capitalism.[101]

The Russian Federation (Russia) has not been a communist state for some time - arguably since the initial signs of the dissolution of the Soviet Union and the fall of the Berlin Wall (November 9, 1989). The cold war and the arms race had certainly taken a toll on the Soviet economy. The "new" Russia represents a somewhat complex transitional economy[102]. The privatization of state resources has led to

---

[100] **Mercantilism**, also called "commercialism," is a system in which a country attempts to amass wealth through trade with other countries, exporting more than it imports and increasing stores of gold and precious metals. It is often considered an outdated system." http://www.vocabulary.com/dictionary/mercantilism. Retrieved 11/18/2015.

**Capitalism** is "an economic system in which investment in and ownership of the means of production, distribution, and exchange of wealth is made and maintained chiefly by private individuals or corporations, especially as contrasted to cooperatively or state-owned means of wealth."
http://dictionary.reference.com/browse/capitalism. Retrieved 11/18/2015.

[101] There is a good concise explanation of "mixed economy" and the reason for it partially explained below:
"Prior to the Great Depression of the 1930s, the United States was primarily a free-market capitalist system and government involvement was minimal. But the massive unemployment and widespread poverty of the Great Depression caused some to believe that capitalism, as an economic system, had failed. John Maynard Keynes revolutionized economic thought and proposed a system of "managed capitalism." As a result of the Keynesian revolution government took a more active role in regulating the economy. This period created a change in the nature of government and the assumption of government's responsibilities. Franklin Delano Roosevelt famously created an economic bill of rights that specified certain rights that were to be afforded to all. These included the right to an education, affordable health care and housing. The government assumed the responsibility to house, feed and educate its citizens."
http://www.uncg.edu/bae/bbt/capitalism/mixed_economy.html. Retrieved 11/25/2015.

[102] "Russia has a high-income mixed economy with state ownership in strategic areas of the economy. Market reforms in the 1990s privatized much of Russian industry and agriculture, with notable exceptions in the energy and defense-related sectors. Russia is unusual among the major economies in the way that it relies on energy revenues to drive growth. The country has an abundance of natural resources, including oil, natural gas and precious metals, which make up a major share of Russia's exports. As of 2012 the oil-and-gas sector

the concentration of wealth into the hands of a few oligarchs. The Putin-facilitated kleptocracy has insured a government that is essentially working in the present and has implicit and explicit aspects of corruption that make any measurable progress or long-range complex planning more difficult.

India is one of the fasting growing capitalist economies and seems to be in position to keep improving due to its relatively young population, investments and savings rates. The government is officially known as the Union Government and is the governing authority of the union of 29 states and seven union territories (the Republic of India). It is a parliamentary system loosely based on the British system but the legal system is strongly based on the English Common and Statutory Law. With its constitutional basis, it strongly resembles several other commonwealth countries but there is a considerably greater complexity inherent in its size, linguistic diversity and multistate composition. India is certainly a capitalist economy and the IMF (International Monetary Fund) has projected extended strong economic performance.

Japan is nominally a capitalist economy since WWII and perhaps much of that is due to the strong intervention of General Douglas MacArthur in the administration of post-war Japan[103]. As in the case of the United States, this nominal capitalism must be clarified by identifying the prime forces in the economy. Certainly, the large "family" corporations (effectively oligarchs) such as Mitsubishi and Toyota that had fully supported Japan's war effort, were quickly and deftly re-tasked by MacArthur to rebuild Japan. There were few if any changes to the Japanese bureaucracy but the major change was a Japanese Constitution that was essentially written by MacArthur's staff.[104] Keeping the emperor in place as a head of state but constitutionally responsible to the will of the people and fundamentally maintaining Japanese culture intact was a masterstroke of policy that insured rapid recovery. However, the fundamental change from a feudal economy to the industrial / technological Japan we now see today was very painful and humiliating for most Japanese. The essential economic point is that Japan is still primarily "ruled" by a handful of very large and entrenched corporate structures. The ownership of prime elements of the economy is highly limited.

---

accounted for 16% of the GDP, 52% of federal budget revenues and over 70% of total exports.[25][26]."
https://en.wikipedia.org/wiki/Economy_of_Russia, retrieved 11/18/2015.

[103] There is much debate about this particular point but there is a very interesting small article by Stanley Weintraub "American Proconsul: How Douglas MacArthur Shaped Postwar Japan", 11/8/2011. http://www.historynet.com/american-proconsul-how-douglas-macarthur-shaped-postwar-japan.htm. Retrieved 11/28/2015. The reference to MacArthur as "Proconsul" harkens back to ancient Rome and is remarkable appropriate in typifying his hands off behavior and his retention of Japan's existing infrastructure and power culture that was very much how the Romans managed conquered territory without causing constant rebellion.

[104] Ibid. "The new constitution had to be ready in a week, in order to forestall any Soviet input. MacArthur's Government Section chief, Brig. Gen. Courtney Whitney, summoned his public administration specialists— some of them lawyers—and announced that they now comprised a constitutional assembly; they would secretly draft the new Japanese constitution, and his three deputies would ensure the document appeared to be of Japanese origin."

The closing of all Japanese nuclear power reactors (nearly 50) after the Fukushima disaster is very telling of the extreme imbalances in power and the lack of an open and viable dialogue about energy production, usage, safety and even responsibility.  Although this has already been addressed, this was a response to public outrage and corporate mismanagement rather than an energy management strategy with any sensible and long-term plan.

The essential questions left, after this brief summary, remain the same and unresolved since there is no defining similarity of ideology or operational approach that would engender any confidence in explaining energy policy or, more basically, the energy use profiles of the five countries.

The next task was to compare the general economic performances of the five countries.  This is a daunting task in-and-of-itself and is continually done by various international bodies such as the IMF (International Monetary Fund) on a regular basis in consultation with the countries themselves.  The following table constructed from IMF "Quarterly Consultations" in 2015 illustrates some of the problems with self-reported economic statistics.  For example, it is inconceivable that Chinese unemployment has been absolutely constant at 4.1% from 2013 through the projected 2016 values?  Also, the multiyear statistics reported for a number of major economic indicators do not indicate any coherent energy policy.

## Compiled IMF Article IV Consultation Data 2015

| Datum | Year | China | USA | Russian Fed. | India | Japan |
|---|---|---|---|---|---|---|
| Real GDP | 2013 | 7.7 | N/A | 1.3 | 4.7 | 1.6 |
| (% annual growth) | 2014 | 7.4 | 2.4 | 0.6 | 5.6 | -0.1 |
| | Projected 2015 | 6.8 | 2.5 | -3.4 | 6.3 | 0.8 |
| | Projected 2016 | 6.3 | 3 | 0.2 | 6.5 | 1.2 |
| Unemployment rate | 2013 | 4.1 | N/A | not listed | not listed | 4 |
| (annual average) | 2014 | 4.1 | 6.2 | not listed | not listed | 3.6 |
| | Projected 2015 | 4.1 | 5.4 | not listed | not listed | 3.7 |
| | Projected 2016 | 4.1 | 5.1 | not listed | not listed | 3.7 |

Part of the emphasis of this book has been to clearly and unequivocally point the finger at the five most responsible parties in the global climate crisis.  Scientifically it is a trivial effort to prove that China, the United States, India, Russia and Japan are the culprits but there is a critical component of any potential solution that is missing – ownership and responsibility for the problem.  One must be struck by the similarity of dealing with the energy addiction of the offending countries and the parallels in treating alcoholism or drug addiction.  There is a strong element of denial.  The fundamental problem is that in dealing with political entities it is incredibly difficult to assess responsibility or have anyone assist in doing so.

Alcohols Anonymous Step 1 is:

> "We admitted we were powerless over alcohol--that our lives had become unmanageable."

For the five major polluters, this can easily be rewritten as:

> "We were powerless over our lust for wealth, power and control -- our environment had become unmanageable."

It may well have been a bit of a waste of time and intellectual space to attempt to examine the governments of the five major global polluters in a moralistic or holistic manner – that is not how they function. It was an attempt to be analytical. Asking the major economic powers to accept responsibility for their actions is unrealistic or, perhaps idealistic.

If the creators of the climate crisis do not accept ownership and responsibility, solutions to the problem become markedly more difficult. It is necessary to consider operational explanations that may provide insights into potential solutions. To this end the next selected statement from the papal encyclical is very telling.

> *Laudato si'* 178. P.63 "A politics concerned with immediate results, supported by consumerist sectors of th*e population, is driven to produce short-term growth. In response to electoral interests, governments are reluctant to upset the public with measures which could affect the level of consumption or create risks for foreign investment. The myopia of power politics delays the inclusion of a far-sighted environmental agenda within the overall agenda of governments."*

**This statement is incredibly incisive** and does something that the church has rarely done – position itself in opposition and open criticism of most ruling governments on the planet. It does have some careful qualifications that might allow the primary offenders (the five leading $CO_2$ producers) to evade responsibility for the climate crisis. It is difficult to carry this criticism of the Vatican forward given the astonishingly positive position that the papacy has taken in openly admonishing all those who would listen for their lack of forethought that borders on operational dishonesty or open insanity.

The essential chord of the criticism is not so much of a particular political system but of the "politics concerned with immediate results". ***This seems to be the fundamental unifying factor in the official behaviors of China, the United States, India, Russia, and Japan***. The "immediate results" are the common factor and that, in turn, is linked directly to economic prosperity and thereby to energy consumption. What is also immediately clear is that any steps that governing bodies any of the offenders might take are potential political suicide. Drastically reducing the carbon footprint will drastically affect any of the economies in question.

*Laudato si'* 194. P.69 *"A technological and economic development which does not leave in its wake a better world and an integrally higher quality of life cannot be considered progress."*

This is another wide-ranging criticism of energy policies that can be applied to most countries or, at minimum to their governments. In a world replete with "bread and circuses, we are inundated with diversion after diversion. It is extremely difficult for most people to determine what are truly significant issues.

If people were polled and asked if they wanted *"a better world and an integrally better quality of life"* they would almost undoubtedly rank this at the top of their list. However, as in all questions and potential answers, the devil is in the details. If one theoretically proceeded to attempt to define the means by which one would achieve this noble and desirable goal it would fast become a pollster's nightmare.

Step one of a theoretical process might be to make some non-partisan, intellectual attempt to define what a "better world" would be.

A better world would be **a world without**:

- war of any kind
- brutality and abuse
- hate and distrust
- grinding poverty and depravation that is not necessary
- religious, ethnic, gender or sexual bias
- excuses and lies for the failure of all of us to be the best humans we can be!

This all seems so right, easy, and natural – so what is the problem? **There is an enormous disconnect between what people say they want and what they actually want.** Added to that disconnect is the gap between what people say and what they ultimately do.

Step two is to determine what "an integrally better quality of life" means. It seems that this is really a two-part question, first defining "quality of life" and then determining the elements by which one would assess a state that was "integrally better". In general, it is very messy when one begins to dissect phrases and sentences in order to get at the concepts.

The only apparent consensus on what "quality of life" is seems to come from oncology literature and not from the ambient population. So, if you are dying of cancer, then there is some attempt to objectify the quality of your rapidly diminishing life.

The emperors of Rome were keenly aware of the power inherent in the control of perception. Roman bread, wine and pork were subsidized[105], losses in foreign campaigns were not common public knowledge, and the complexity of running a widespread and culturally diverse empire was virtually never a topic of discussion. In much the same way, modern governments of nations or, at minimum, the politicians associated with them, have no vested interest in tackling large (international) and technically difficult problems.

Political campaigns, especially in the United States, often focus deliberately on hot-button and divisive issues such as immigration, gun control, abortion rights, and global terrorism. These issues are either of no real economic import or cannot be controlled in any practical way by politicians.

A final quotation is most deeply telling and analytical – it is a call to action, a call to spiritual and moral awareness, a call to responsibility and to the best part of our humanity.

> *Laudato si'* 217. P. 76 *"The external deserts in the world are growing, because the internal deserts have become so vast". [152] For this reason, the ecological crisis is also a summons to profound interior conversion."*

As a non-religious person, I am still moved and motivated to take whatever actions are necessary to affect the changes that are essential. We must all look deeply within ourselves to root out complacency, negativity, and the feelings that we cannot make a difference. I flatter myself in believing that I might emulate James Henry Leigh Hunt's "Abu Ben Adhem"[106] who is not in the books of the lord but tells the visiting angel to "Write me as one who loves his fellow men".

What do "people of faith" and non-religious people have in common that could motivate us and send us along the same path – potentially toward a solution to a large and significant problem? Perhaps the answer is as simple (and as complex) as our world itself. There is nothing more important than the environment that we live in! From the first picture of the Earth from the moon it has been abundantly clear to anyone who could think - the planet is beautiful, rare and fragile. We may not be unique but we are rare and for that reason alone **it is important to understand our place in the universe and our role in the world – that of intelligent stewardship not wanton exploitation.**

---

[105] The *cura annonae* ("care of the grain supply" – literally "care of provisions") as the empire grew and grain had to be imported in ever larger quantities to feed the growing population. During the reign of Septimius Severus, olive oil was added to the distribution. During the reign of Aurelian, a major reorganization occurred and it appears that he ceased to distribute free grain; instead, he issued free bread, and added salt, pork and wine to the dole, which was provided free or at a reduced cost. These measures were continued by successive regimes. Southern, Pat, *The Roman Empire from Severus to Constantine* (2004), pg. 326. It is clear that these subsidies were a major drain on the treasury.

[106] James Henry Leigh Hunt (1784-1859). The poem is very short and well worth reading. See: http://www.poetryfoundation.org/poem/173698. Retrieved 09/09/2015.

## WHAT CAN "PEOPLE OF FAITH" DO?

People of faith can do much more than pray. That is certainly incredibly apparent in the political muscle that they have managed to wield in the United States. It is exactly this type of political muscle that could be used to pressure government(s) to behave more responsibly in terms of environmental issues of global significance.

A recent article in the National Review[107] states that:

> "A recent study reveals that faith-based groups contribute more to the U.S. economy than the top ten technology companies — including Google, Apple, and Amazon combined — producing a total revenue of $1.2 trillion each year. If this figure were put in terms of GDP, U.S. religion, as defined by researchers, would be the 15th largest national economy in the world."

This represents an enormous power if it were wielded in a calculated and consistent manner. One could write an entire book on how this power could be marshalled and used.

This "power of faith" is not limited to Christians - Muslims represent another large group. The Pew Research Center produced a "demographic study – based on analysis of more than 2,500 censuses, surveys and population registers – finds 2.2 billion Christians (32% of the world's population), 1.6 billion Muslims (23%), 1 billion Hindus (15%), nearly 500 million Buddhists (7%) and 14 million Jews (0.2%) around the world as of 2010."[108]

If indeed people of all faiths could agree on the critical nature of environmental stewardship this could be a game changing force for the betterment of the planet. How? Miraculous leadership ... no pun intended whatever.

---

[107] Desanctis, Alexandra. "Faith-based groups contribute enormously to American society and the U.S. economy." The National Review; on-line September 22, 2016 4:00 AM. http://www.nationalreview.com/article/440269/religion-economy-faith-based-groups-contribute-us-economy. Retrieved 03/23/2017.

[108] Pew Research Center; The Pew Forum on Religion & Public Life; *The Global Religious Landscape: A Report on the Size and Distribution of the World's Major Religious Groups as of 2010*, December 2012. PDF available at http://www.pewforum.org/2012/12/18/global-religious-landscape-exec/. Retrieved 06/13/2017.

# WHAT SHOULD SCIENTISTS DO?

"If you are not a part of the solution, you are a part of the problem."

Eldridge Cleaver (American author and political activist – early leader of the Black Panther Party, August 31, 1933 - May 1, 1998)

**SCIENTISTS** … MAN/WOMAN UP! We cannot sit back behind a curtain of impartiality or the illusion that we are too small or ineffectual to make a difference - we must act! The words of the Pope may be from a higher authority but given that many scientists are not people of religious faith there are perhaps equally as powerful imperatives:

- academic and intellectual honesty
- faith in the methodology and operational integrity of science – that the crisis is imminent!
- recognition of the irrefutable proof that the political approach to solving global climate problems has been utterly ineffective even when the crisis is apparent.

Scientists control more than they think on an international scale. They can:

- demand responsive and responsible energy policy from their colleagues and the governments that employ them
- aggressively attack rhetoric and bad science that contributes to poor energy policy choices
- run for public office as responsible characters in what is generally a 3-ring-circus
- form an International Consortium for credible data analysis and creative solutions to global warming. The Intergovernmental Panel on Climate Change (IPCC) is a UN organization that has taken steps in that direction but the message seems to be somewhat unheard.[109]
- force universities and corporations in which they are staff to take up the gauntlet with their respective intransigent governments
- design and deliver POPULAR (designed for lay people) information sources and presentations that begin the very difficult process of educating people with verifiable data regarding their options and the paths towards a sustainable climate and sensible energy policy.

**DON'T march for science … DO science … BE science … have the courage to speak out at EVERY OPPORTUNITY.**

---

[109] The IPCC and Al Gore jointly won the Nobel Prize in 2007 "for their efforts to build up and disseminate greater knowledge about man-made climate change, and to lay the foundations for the measures that are needed to counteract such change". http://www.nobelprize.org/nobel_prizes/peace/laureates/2007/. Retrieved 07/26/2017.

## Fundamental Principles (Fondest Wishes)

Energy should not be a weapon!  The economic and social development of the globe should be the goal of all people.

Politicians MUST be neutralized or heavily coached regarding all technical discussions and formulation of solutions to global warming! ***This alone is a monumental if not impossible task.*** Perhaps politicians could be reeducated by their own constituents to be more responsible and considerably better informed on critical issues.

There may be a ray of hope in the recent elections in both France and in England.  In France, Emmanuel Macron of En Marche! has fashioned a new powerful centrist party that has the possibility of steering a course that the people really want.  The old, polarized and opposing party positions are no longer popular with voters.  Perhaps the result of the recent early election called by the conservative prime minister, Theresa May, in the UK provides a less clear reading of the preferences of the electorate but her very narrow win leads to a "hung parliament" with no clear majority and it will make the negotiations to leave the European Union (Brexit) very much more difficult.

Marcon's victory in France is very heartening when set against the nationalist movement in Britain and the Trump administration's heightened isolationist rhetoric.  To solve a global problem we need global leadership, cooperation and common goals not just for nations but for the planet itself.  In this sense, the Pope was correct in invoking mankind's obligation to provide a true stewardship of nature.

# PLANS TO MOVE FORWARD - SOLUTIONS

*"If you do not change direction, you may end up where you are heading."*

Lao Tzu [Laozi] (Chinese Philosopher and Poet, 6th Century BCE)

Change is a difficult concept for many individuals and a much more difficult one for organizations and governments. Stability and familiarity is what many seek in a world that seems too chaotic, noisy, and inhospitable. However daunting change is, it is the only way forward, it is the route of progress around particularly difficult impediments. We can do as we have been doing and be swept away with the awesome tide of climate change or we can make fundamental changes to our energy consumption and production.

Changing direction is what we ... collectively ... globally must do. These words rarely appear together. Up until this time we have lived in a world composed of independent nations that work diligently in their own interests. The few international organizations like the United Nations have often been deliberately sidelined on many important questions due to the politics between individual nations and blocks of opponents. The UN Security Council still often behaves as if the Cold War was still on. **Perhaps a new and urgent task of the United Nations (at a higher level than the IPCC) is to be united against a real global threat!**

As Lao Tzu has suggested ... we must change direction ... NOW and EVERY DAY FORWARD.

**The first, major element of that change is to help most people to change their negative impression of nuclear power.** We have had nuclear power reactors for decades and they have, by-and-large, been incredibly efficient in providing power in an enormously cleaner manner than any fossil fuel. However, public perception of nuclear power is exceptionally low for NO INTELLIGENT OR LOGICAL REASON. Therefore, a massive educational campaign that must be entered into as soon as possible.

One of the first and most important steps is to re-educate people. It is often extremely difficult to convince people that what they believed is fundamentally flawed but that is exactly what is required. The next sections on nuclear power are a brave first step in trying to get people to rethink their stance on nuclear power.

I make no apologies for the asides that I believe are critical to formulate a more comprehensive understanding of why we should move forward and perhaps even how we could do that.

# NUCLEAR POWER – A BACKGOUND AND SOME LESSONS

In an earlier section "PERCEPTION VERSUS PERSPECTIVE" a graphic of a poll of "Global public support for energy sources" May 2011, was presented with a short, pointed and slightly quizzical discussion as to how coal could remotely be conceived as a better choice than nuclear. This is the only graphic that will be duplicated in this book because of its central importance to one of the main solutions to global warming.

## Global public support for energy sources

"Please indicate whether you strongly support, somewhat support, somewhat oppose, or strongly oppose each way of producing energy"

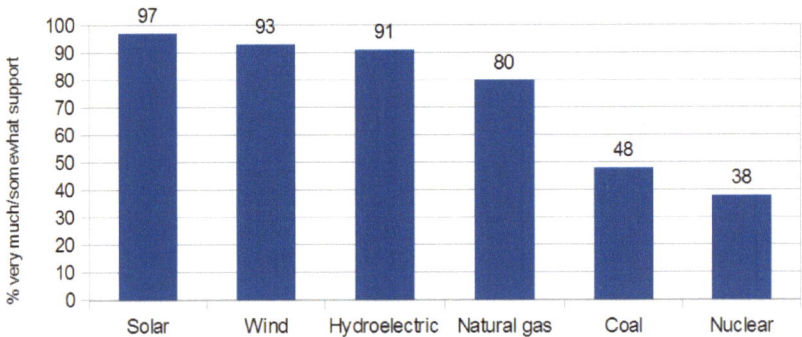

Source: Ipsos, May 2011

This graphic should alarm any scientist, or thinking non-scientist, how improperly skewed public opinion can be. The consequence of that is lack of financial and political support for what should be done. The following sections on Fusion and Fission are the very simplest start or suggestion of what would have to be done to begin a change in perception towards the development of a fundamental understanding of energy issues.

# FUSION

This entire section was never really intended to be included in this book but it is critically important to understand the hurdles and bumps in the road towards "Solutions". One of the greatest hurdles was presented above – "Global public support for energy sources". I do not wish to be perceived as biased or falsely pro-nuclear but I will invoke the intellectual comradeship of perhaps one of the finest minds the world has ever produced ... Stephen Hawking.

> *"I would like nuclear fusion to become a practical power source. It would provide an inexhaustible supply of energy, without pollution or global warming."*
>
> Stephen Hawking (British Physicist, In *Time* Magazine, "10 Questions for Stephan Hawking", Monday November 15, 2010. The question was: Which scientific discovery or advance would you like to see in your lifetime? —*Luca Zanzi, ALLSTON, MASS.")*

## A COSMIC ASIDE ... "WE ARE STARDUST HARVESTING STARLIGHT"[110]

> *"It is far better to grasp the universe as it really is than to persist in delusion, however satisfying and reassuring."*
>
> Carl Sagan (American Scientist, November 9, 1934 - December 20, 1996)

An effective and scientifically based energy reeducation campaign is essential to bring public opinion around to see that nuclear is one of the best tracks towards positive climate change and global energy use. With nuclear weapons and disasters like Fukishima in peoples' minds and persistently in the news it is extremely difficult to get people to focus on real facts (not the "alternative facts" that seem to be prevalent in the Trump Universe).

So ... please bear with this lengthy aside which should provide the logical and technical background for pushing forward with one of a number of potential solutions that will be discussed later and should all be pursued in parallel in as much as they will not interfere with the ultimate goals – reining in $CO_2$ and reducing climate warming.

---

[110] In his television show the *Cosmos: A Personal Voyage*, Carl Sagan often waxed poetic about our place in the universe but one of the most fundamentally correct things he said, "we are star stuff harvesting starlight".

When we move out of doors few of us take the very few seconds it takes to look up into the sky and to ponder something fundamental and universal – **we are an intimate part of a nuclear family – the fusion family!**

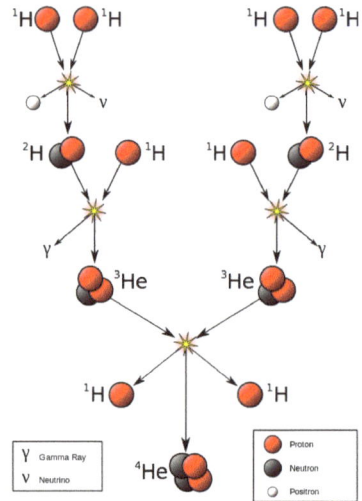

Whether our few seconds of contemplation is during the day or at night it would be hard to miss the sun or the approximately 100-400 billion stars (suns) in the Milky Way. Every one of these points of light is a fusion reactor. Our sun, as do all stars, fuses hydrogen to make helium and vast quantities of energy.

There are many reactions that give rise to fusion in the sun but the basic idea or reaction is as seen in the diagram. The diagram[111] shows the chain that must occur to produce helium (He) because neutrons are required to continue fusion reactions to higher elements.

The proton–proton reaction, as illustrated above, dominates in stars the size of our sun or smaller.

If we wish to understand the energy involved in fusion we need the help of good old Albert Einstein to make a rough calculation from $E = mc^2$. **For 1 gram of matter (1/453 lb.) the energy in ergs** = 1 gram x (30,000,000,000 cm/sec) x (30,000,000,000 cm/sec) = 900,000,000,000,000,000,000 ergs. In terms of an understandable energy output this would be the lighting in a small town for 1 year. This is indeed enormous!

The simple and undeniable fact is that the detectable light in the universe is caused by fusion. An equivalent but often mis-communicated fact is that all the matter except hydrogen and some originally existing helium is a result of fusion. Everything we see and can sense is a direct result of a nuclear process (usually fusion).

Our sun is quite young (about 4.567 billion years old[112]) relative to the beginning of star formation at the time of the "Big Bang", approximately 13.798 ± 0.037 billion years ago.[113] It formed due to the gravitational collapse of a pre-existing dense molecular cloud.

---

[111] http://en.wikipedia.org/wiki/en:User:Borb?rdfrom=commons:User:Borb.
[112] Connelly, James N.; Bizzarro, Martin; Krot, Alexander N.; Nordlund, Åke; Wielandt, Daniel; Ivanova, Marina A. (2

Our own solar system is not the origin of the material that makes up the bulk of the earth. The earth has a very heavy liquid core of iron (atomic no. 26) and nickel (atomic no. 28). These elements are not produced in any significant quantity by our sun which has the spectral class G2V. The heavier elements in our solar system came from the same molecular cloud that formed our sun.

Spectral classification of a star (any sun) is generated by describing the ionization of its photosphere and provides an objective measure of the photosphere temperature. The absorption spectra of various elements indicate temperatures sufficient for ionization of the element. Stars are classified using the letters O, B, A, F, G, K, and M ("Oh Boy An F Grade Kills Me"), where O stars are the hottest and the letter sequence indicates successively cooler stars. A useful tool is to link observed color to the class and it is generally said that, O stars are "blue", B stars are "blue-white", A stars are "white", F stars are "yellow-white", G stars are "yellow", K stars are "orange", and M stars are "red". No-one should take the classification of our sun as a "yellow dwarf" personally.

The current non-alphabetical scheme developed from an earlier scheme using all letters from A to O; the original letters were retained but the star classes were re-ordered in the current temperature order when the connection between the stars' class and temperatures became scientifically apparent.

The Morgan-Keenan system uses the spectrum letter modified by a number from 0 to 9 indicating tenths of the range between two star classes. Luminosity class is expressed by Roman numerals I, II, III, IV and V, indicating the width of certain absorption lines in the star's spectrum. Class I are called supergiants, class III simply giants and class V, either dwarfs or, more properly, main-sequence stars.

So, good old Sol, our sun, is a G2V yellow dwarf star. That means it is yellow and 2 (2/10's) of its way to orange and it is V or a main-sequence star. The problem with Sol is that it could not have produced the heavy elements that formed the terrestrial planets and asteroids that litter our solar system.

So where did all of the material of our solar system come from? We did evolve from a molecular cloud but what was its origin? Recently accepted stellar formation models indicate that high-mass O and B stars that are hot and very short-lived produce heavy elements at a rapid rate. There is more than a little curve that nature throws us here and that is fusion, by the proton-proton reaction (that is one

November 2012). "The Absolute Chronology and Thermal Processing of Solids in the Solar Protoplanetary Disk". Science 338 (6107): 651–655.

[113] Planck collaboration (2013). "Planck 2013 results. XVI. Cosmological parameters". Submitted to Astronomy & Astrophysics. arXiv:1303.5076. Bibcode:2014A&A...571A..16P. doi:10.1051/0004-6361/201321591.

hydrogen plus one hydrogen), proceeds predominantly in stars below about 1.3 times the mass of our sun but above that mass a very different process predominates and that is the CNO cycle[114].

| | | |
|---|---|---|
| ⬤ Proton | γ | Gamma Ray |
| ⬤ Neutron | ν | Neutrino |
| ◯ Positron | | |

This is the primary cycle for the very large amount fusion that occurs in heavier stars. The O and B stars rapidly proceed through their short lives (as little as a few million years) to become supernovae. These massive explosive events produce the building blocks for new solar systems with heavier element distributions. Due to the short life-span of such massive stars there have been many such life-cycles since Big Bang (approximately 13.7 billion years ago). They have provided the base materials of terrestrial planets as well as the carbon that is, at least for creatures like us, an essential material of life as we know it.

Carl Sagan often waxed poetic about our place in the universe but one of the most fundamentally correct things he said was "we are star stuff harvesting starlight". We are indeed the indirect product of nuclear fusion. Every carbon atom in every person was forged in the nuclear furnace. *We, all life, are the result of nuclear reactions*. There is no other explanation that bears up to any scrutiny at all. Science is not convenient or comfortable but it is often correct. **So "NUCLEAR" is not bad! It is GOOD!**

Since we are all nuclear material and our planet is composed of material made in the fusion cycles of ancient stars how is it logical or intelligent that so many people are "anti-nuclear"?

---

[114] The CNO cycle is the Carbon, Nitrogen, Oxygen fusion cycle that is prevalent stars with mass in excess of 1.3 times that of Sol. Graphic by Borb, CC BY-SA 3.0, https://commons.wikimedia.org/w/index.php?curid=691758.

What does it even mean to be anti-nuclear?  Are you:

1) antithetical to the creation and use of nuclear weapons (either fission or fusion)
2) against the peaceful use of nuclear power
3) horrified that isotopes made in reactors are used routinely in medicine and often save lives
4) vehemently against the primary source of all energy in the universe
5) hateful of a tan or dislike sunlight, starlight or moonlight?

For a scientist, it is very difficult to pose these questions in a less offensive form.  Science, thank the stars, is not about making people feel better about their personal beliefs and faith.  It is a mode of seeing the universe that is fully subject to independent scrutiny, discernible proof and refutation.  Unlike politics and most religions, science has rules of evidence and procedures to replicate and verify results.

The facts are clear.  The universe is, in part, a nuclear furnace.  No opinion, no debate, no useful discussion is possible once we have done our basic homework.  So, is there any purpose to the homework?  The most common speculation one can come to, based on the available facts, is that "nature really can't be wrong"[115] in the proliferation of such a universal source of power.  Fusion is what is natural or it is a least what nature tends to do a great deal of.

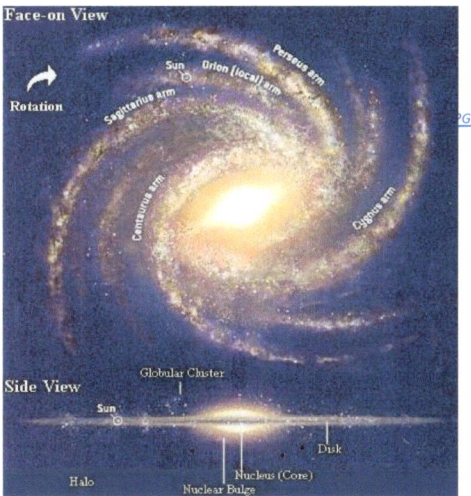

There are some other fundamental issues that require attention.  There are several natural consequences of fusion such as the building of planets and the formation of living things.  It is important to think clearly about the fusion engine that has brought life to this planet.  It is not plausible that life, and even intelligent life exists only here.  No matter how special life seems to be, the sheer vastness of the universe and the numbers of stars statistically demands that there is life in many other places.  A simplified diagram of the Milky Way with the relative position of our sun is shown in the diagram.

---

[115] "Nature really can't be wrong" is probably one of the most annoying pseudo-truisms and anthropomorphisms ever proffered for linguistic and intellectual dissection.  Nature is not a being, does not have a will and likely does not have a definable purpose – nature simply is.  Our striving to understand it and our place in the universe speaks volumes about us and our amusing self-obsession and not about nature itself.

We really do not have to look beyond the Milky Way Galaxy. Regardless of whether one accepts the very conservative lower estimate of 200 billion stars or the more likely 400 billion stars in our immediate neighborhood the number of stars is staggering. Why is there such a difference in the estimates? Well there is not a simple single answer but the first problem is that we really can't "see" the Milky Way in the manner in which we see other galaxies. Our position within the outer arms of the galactic disk does not allow us to see many areas on the other side of the disk. There are too many stars and obscuring nebulae (molecular clouds and dust) concentrated towards the center of the galactic disk.

Besides the very simple physical problems implied above there are deeper observational problems. With the steady progress in cosmology and the refinement of stellar evolution models paired with observational evidence it seems that there are very many small low-luminosity stars (brown and red dwarfs) that are very difficult to observe even over relatively small intragalactic distances.

Although we almost certainly have many neighbors in our own galaxy we must begin to accept another rather cruel reality - we may never see or even communicate with any of them.

The Kepler Space Telescope (http://kepler.nasa.gov/) was launched on March 7, 2009 to look for exoplanets, our nearest neighbors, in our local Orion Spur of the Milky Way.

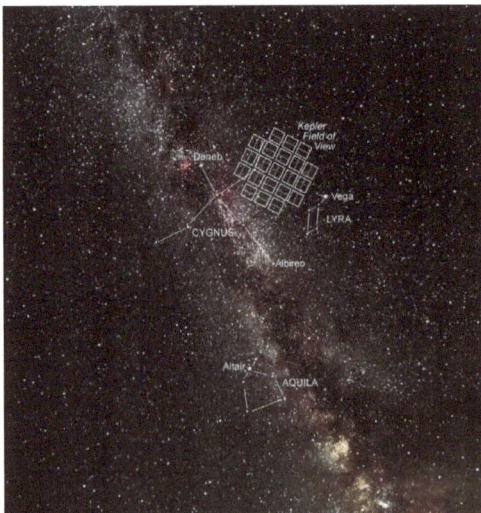

Kepler's field of view is shown in the image (Kepler Mission Star Field - An image by Carter Roberts of the Eastbay Astronomical Society in Oakland, CA). Some results are shown below:

| | |
|---|---|
| *Eclipsing Binary Stars* | *2165* |
| *Planetary Candidates* | *4796* |
| *Confirmed Planets* | *2330* |
| *Confirmed in Habitable Zone* | *~12* |

*Data as of 10/11/2016*[116]

---

[116] http://kepler.nasa.gov/.

The area indicated in the image was found to have greater than 160,000 viewable stars and after several years, the findings were stunning. One must remember that this represents only a tiny local section of our galaxy. The findings have drastically changed notions of the frequency and distributions of planets within our galaxy. We have a large number of near neighbor planets, some habitable. What do we make of that and what does it have to do with fusion and this book?

The two fundamental causes for our isolation and silent loneliness are the distances involved (and, hence the time frame) and our old friend from classical physics – the inverse square law. The first of these is important to consider. Our nearest stellar neighbor is Proxima Centauri (V645 Centaurus or Alpha Centauri C, part of a trinary star system) at 4.24 light-years distance. Voyager 1 (launched on September 5, 1977) has an approximate speed of 17.05 kilometers per second, making it the fastest outward bound spacecraft ever[117] launched (at that time). Even at this velocity it has taken Voyager 1, thirty-eight (38) years to get to interstellar space. Voyager 2 launched on August 20, 1977 and has made its way to the heliosheath[118] (position as of May 2015). The two Voyagers are the only probes that have been sent into interstellar space and that is by merit of their very sturdy design and the fact that NASA decided to drastically extend their mission lives in order to do the first investigations outside of the solar system. It is strangely riveting to go onto the Voyager website and to look at a real-time odometer for both craft http://voyager.jpl.nasa.gov/where/index.html).

At 11:10 PST on 02/21/2017, as this section was re-written, Voyager 1 was 20.604 billion kilometers from earth with a roundtrip light time of 38hr, 14min, 09sec. Voyager 2 was 16.997 billion kilometers away with a roundtrip light time of 31hr, 39min, 49sec.[119] It is important to put the times and distances into perspective because the roundtrip light time to the moon is about 2.6 seconds. By just doing some simple division of the roundtrip light times, Voyager 1 is 52,548 times further away from earth than the moon and Voyager 2 is 43,173 times further away from earth than the moon. It is a long way to interstellar space from earth and the Voyager craft are very far away indeed, yet these distances are insignificant relative to interstellar distances.

> "Voyager 2, which was launched in August 1977 and is still functioning despite the mind-boggling distances and numbers involved, has set course for Sirius – the brightest star in the sky.
>
> Astonishingly, even travelling at around 40,000mph it will take 296,000 years to reach Sirius."[120]

---

[117] Proxima Centauri is a red dwarf variable star of spectral class M5.5 Ve and is not visible to the naked eye. The constellation Centaurus is visible in southern skies and is one of the largest constellations.

[118] The heliosheath is the region of the heliosphere beyond the termination shock. Here the [solar] wind is slowed, compressed and made turbulent by its interaction with the interstellar medium. Its distance from the Sun is approximately 80 to 100 astronomical units (AU) at its closest point. Definition adapted from in Wikipedia.

[119] The Voyager real time odometer can be found at http://voyager.jpl.nasa.gov/mission/weekly-reports/. Retrieved 02/21/2017 but be aware that NASA routinely archives and deactivates them depending on staff commitments.

[120] http://www.express.co.uk/news/world/567957/NASA-s-Voyager-2-sets-course-for-star-Sirius-by-time-it-arrives-human-race-will-be-dead. Retrieved 02/21/2017.

The very fine Upper Paleolithic paintings in the Lascaux caves in the Dordogne department in France have been dated to 17,300 years old – we were cave men! The approximate date for the period of maximum glaciation in the last ice age is about 22,000 years ago. Our Neanderthal relatives became extinct around 30,000 years ago. When your nearest neighbor star is 4.243 light years away (40,114,403,776,136 km ~ 40 trillion km) you live in an awfully large neighborhood and the likelihood of casual visitors seems as remote as the neighborhood is vast. Even if we magically could travel ten times faster than our fastest spacecraft it would still take about 7,400 years to go to borrow a cup of sugar (double that time if you wish to return to do the baking). So just the distance is a great physical and temporal stumbling block. Although many of us – even the most hard-core scientists – would love to believe in E.T.[121] it is just not very scientific or realistic to expect him/her/it to visit.

The second problem or issue in our isolation is the silence or perhaps, more accurately, the overwhelming reality that we are functionally mute. The inverse square law (as graphically shown below) makes any attempt at real communication more than a little laughable. It does not take very much time/distance to make any signal reduce to levels of background noise.

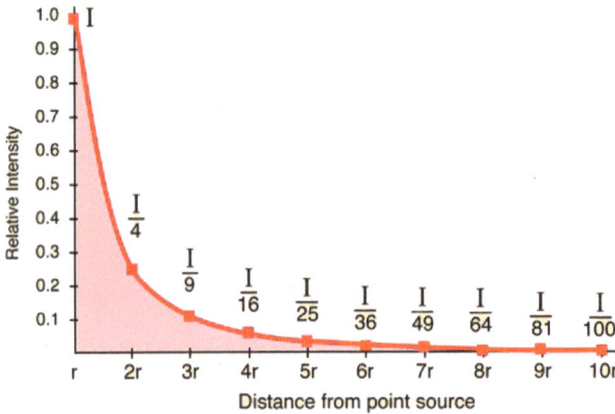

At 10 radii in a sphere we are at 1/100 of the intensity. At 100 radii 1/10,000. Even our loudest boozy global party would not be audible or visible at our nearest stellar neighbor.

*Image retrieved 11/12/2016 from http://hyperphysics.phy-str.gsu.edu/hbase/Acoustic/invsqs.html*

---

[121] Steven Spielberg's marvelous naïvety is admirable but if we were to receive an alien visitor my ideal choice would be "Paul" as depicted in the 2011 film written by Simon Pegg and Nick Frost. He is much more jaded and "earthy". A creature that would not make me too embarrassed to be the flawed human that I am lucky to be.

We are not alone in the universe but we are very alone here, on earth, in the Orion Spur of the Milky Way galaxy. As human beings, we must embrace the wonder of who we are, where we are and when we are. This is very much the message in one of the only movies that attempts to deal with the problems of communication in the universe – *Contact*[122] (1997) starring Jodie Foster. The participation of Carl Sagan[123] in the seminal role of writer of the book as well as writer for the film story is very apparent in the nature of the contact that is suggested. It is essentially by way of the theoretical Einstein-Rosen bridge[124] or wormhole between dimensions or universes. Basically, normal relativity would not matter for time and distance if there were a dimensional bridge. Stephen Hawking has discussed "Space and Time Warps" in a lecture on his website http://www.hawking.org.uk/space-and-time-warps.html. In typical fashion, he would not look favorably on such a "violent" means of intelligent communication although it may be the only one we could hope for in such a vast place as our universe. Then, what can we do?

**Science fiction may inspire us but science will inform us**. We must embrace the science that has indeed distinguished us from other animals and use sound scientific reasoning to manage ourselves and our planet. We must accept the responsibility for what happens to our world. We don't have the Hollywood or science fiction option of finding another planet to replace this one and we are not going to get help from aliens so we better make the best of the planet we live on!

**Without alien intervention what do we need to do to manage our planet?** At first glance, the issues seem too complicated and overwhelming. It may seem heretical to suggest but they are not. We need to manage our resources. What are those resources? The list is pretty short – land, water, air, and energy. If this sounds like the ancient four "elements", earth, water, air and fire, it essentially is the same. These basic entities and concerns have not materially changed over the last few millennia. What has changed is our ability to understand them and to manipulate them. The most critical of the fundamental elements is energy. The production and consumption of energy is a major preoccupation of the developed and developing world and its proper management will be a mark of our own evolution or self-inflicted injuries.

---

[122] http://www.imdb.com/title/tt0118884/. The IMDb credits list Robert Zemeckis (director) and Writers: Carl Sagan (novel), Carl Sagan (story), Ann Druyan (story), James V. Hart (screenplay) and Michael Goldenberg (screenplay).

[123] Carl Sagan died December 20, 1996, before completion of the film.

[124] "Einstein Attacks Quantum Theory" read the *New York Times* headline of May 4, 1935. The article continued: *"Professor Albert Einstein will attack science's important theory of quantum mechanics, a theory of which he was a sort of grandfather. He concludes that while it is "correct" it is not "complete." With two colleagues at the Institute for Advanced Study here, the noted scientist is about to report to the American Physical Society what is wrong with the theory of quantum mechanics. The quantum theory with which science predicts with some success inter-atomic happenings does not meet the requirements for a satisfactory physical theory, Professor Einstein will report in a joint paper with Dr. Boris Podolsky and Dr. N. Rosen."*

This is a post on the Princeton Institute for Advanced Studies website that properly introduces the concept (https://www.ias.edu/articles/einstein-rosen-bridge).

It is of critical importance that there be some consensus (at least in places where education is available and encouraged) that science and technology are the fundamental engines of human advancement – **not politics or religion.**

So why is energy policy not directed by sound judgment and scientific thinking?  Would there be a climate crisis had the responsible scientific approach to acid rain and climate change been taken seriously in the late 1970's?  Perhaps that question is disingenuous and self-serving.  There is much to discover about fundamentals before anyone can ponder some of the less resolvable aspects of problems that we face – the human element and the much less understandable political element.

This book tries to take a more global and fundamental approach to energy and the environment.  From the outset there is no attempt to pretend that there can be a rational discussion with people who insist on being irrational.  However, *we must begin to be rational about problems we share as human beings and attempt to develop a framework for problem resolution and development of intelligent and scientifically credible public policy.*

# FUSION – ON EARTH

After introducing fusion as the model for power production in the universe it is more than humbling to have to admit that our attempts at fusion on earth have been far less than successful to this point.

It should be the simplest possible thing to do – join the two simplest elemental forms in the universe. Each hydrogen atom ($^1$H) is just one proton and one electron and the target that one wishes to create is one helium atom – two protons two neutrons and two electrons. Well, there is a problem right away. Where do we get the two extra neutrons?

We really are seeing first, the creation of a heavy hydrogen isotope $^2$H (deuterium), then a second fusion with $^1$H (protium). This produces an isotope of helium ($^3$He) which is one neutron short of what is the most stable[125] isotope $^4$He, which is created with yet another fusion cycle.

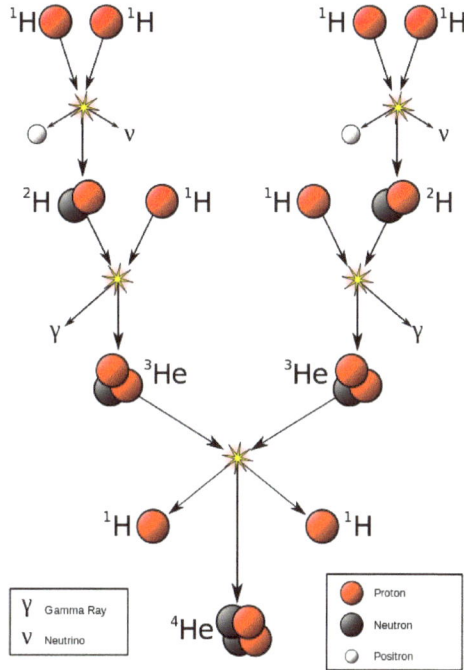

So, I lied – it is not the simplest thing to do! However, there are fundamental aspects of the entire process of fusion that we must explore before we can find solutions to overcoming those problems.

In a purely statistical sense, the possibility of $^1$H meeting $^1$H in a stellar core is quite high since it represents a large fraction of what is there. Our own sun, Sol, has a core density of up to 150 g/cm3 (about 150 times the density of water) and a temperature of close to 15.7 million Kelvin degrees (K). Under those conditions fusion can take place.

---

[125] It is somewhat problematic to discuss the "most stable isotope", perhaps what I should have said is the most familiar isotope with the specific meaning; that is, it is the most stable isotope for humans living on earth who may have done some basic chemistry and who do not live inside a star.

This meeting of two hydrogen nuclei is the start of the proton-proton reaction (pp-reaction)[126] that is the major way of making energy in a less massive star and one of the paths that lead to heavier elements. The second step is also a statistically high probability because all a $^2$H nucleus has to do is meet a $^1$H nucleus (basically $^1$H is everywhere as it is the most abundant isotope in the sun and the universe).

Now step three is that $^3$He and $^3$He will meet, and produce $^4$He. In a purely mathematical sense this is much less likely due to the relative scarcity of $^3$He. But statistics slides in to make the logical save – we have to envision a stellar core with very high energy, very high density (proximity), and very high collision rate. Therefore, even the less likely events are occurring regularly.

It is surprisingly difficult to come up with a good analogy in normal human experience but perhaps some of us have been in a pool hall or a den with a pool table. The table has only red balls on it and when 2 red balls collide very hard they partially stick together to form a yellow one. Initially there will only be one yellow ball but soon there will be another and another because there are so many red ones hitting each other. These could fall apart to become 2 individual red balls but some don't, so that very soon, within the confines of the pool table, 2 yellow balls will meet with sufficient force to produce a blue ball. Of course, this analogy implies that we had too much to drink or a psychotropic drug. The reality of the universe is often far outside of our personal experience – our minds must take us where the limitations of our own physical reality and simple pool hall analogy cannot.

The basic analogy is enough to suggest that eventually we build larger and larger atoms (heavier elements) and that they will generally be less abundant the heavier they are – of course there are some very interesting exceptions, like iron[127] but looking into that would divert us even further away from the fundamental point of this book.

## Early "Experiments" with Fusion

It is kinder to call the first hydrogen bomb an experiment than what it really became – part of a problem. The problem is that there is a widespread popular belief that nuclear power is strongly related to nuclear weapons. As with many technologies throughout history, major advancements have been made in the cauldron of war but that does not always remain true. The issue of power versus weapons will be considered more carefully in the next chapter. Introducing weapons into the discussion is appropriate and necessary since they really came first.

---

[126] There are several branches of the p-p reactions sequence but they do not affect the basic points being made and are best left to an introductory nuclear physics / chemistry class than the central point of this book.

[127] In smaller stars where the CNO cycle does not predominate the heaviest element or core nuclei that are stable is nickel which decays to iron.

The first H-bomb (codename "Ivy Mike"[128]) was detonated on November 1, 1952, on Eniwetok atoll in the Pacific. This was well after the A-bombs (fission weapons) "Little Boy" were dropped on Hiroshima (August 6, 1945) and "Fat Man" on Nagasaki (August 9, 1945). The H-bomb and fusion are presented first in this account since fusion preceded fission in the universe and in the creation of matter[129] - without fusion there can be no fission.

The H-bomb was a direct result of the cold war and the nuclear arms race with the Soviet Union. Following the successful Soviet detonation of a fission weapon in 1949[130], the United States accelerated its program to develop the next stage, a much more powerful thermonuclear or H-bomb. The new weapon, "Ivy Mike" was approximately 693 times more powerful than "Little Boy".

J. Robert Oppenheimer, one of the fathers of the atomic bomb, opposed development of the H-bomb arguing that it would only speed up the arms race. The Soviet Union exploded their first thermonuclear device on August 12, 1953, less than a year after "Ivy Mike" and the arms race was on full throttle. To scale the enormity of the destructive potential of the 10.4 megaton consider "Ivy Mike" compared to all of the bombs dropped in WWII – approximately 2.8 megatons[131]. Perhaps Oppenheimer saw most clearly that this was the kind of race that no-one was going to win.

The development of a functioning fusion reactor (Stephen Hawkings' wish) would indeed be a major milestone for humanity but it still seems to be a number of years away. It takes a great deal of energy to make a fusion reaction occur and fission is a much more facile process that occurs without the need for the massive gravity/density/temperature present in a stellar core.

---

[128] "Ivy Mike" was 10.4 megatons (http://en.wikipedia.org/wiki/Ivy_Mike) versus a mere 15 kilotons for the "Little Boy" A-bomb dropped on Hiroshima (http://en.wikipedia.org/wiki/Little_Boy).

[129] At the instant of the Big Bang (13.82 billion years ago according to measurements made by ESA's Planck Satellite launched May 14, 2009) the composition of matter in the newly cooled universe (about 300,000 years after Big Bang) was approximately 73% hydrogen and 25% helium with 2% being everything else. See a basic discussion in http://hyperphysics.phy-astr.gsu.edu/hbase/astro/hydhel.html.

[130] On 29 August 1949, the Soviet Union conducted its first nuclear test, code-named 'RDS-1', at the Semipalatinsk test site in modern-day Kazakhstan. The device had a yield of 22 kilotons of TNT.

[131] Total estimated tonnage dropped in WWII – 1,463,432 (USAF) + 1,307,117 (RAF) = 2,770,540 tons (2.8 megatons) retrieved 06/13/2015 from https://en.wikipedia.org/wiki/Strategic_bombing_during_World_War_II.

As mentioned above, fusion is more difficult to achieve than fission and the fundamental reason is that one has to overcome the very high electromagnetic force that intrinsically keeps atoms apart.  If there were no resistance to uniting, all matter would melt together in an indistinguishable singular blob (or something like that).  Atoms are balanced structures of attraction and repulsion.  The more naïve view of the atom as negative electrons orbiting a positive nucleus will be sufficient to start a simplified discussion of the problem of fusing two hydrogen atoms.

It is not the negatively charged outer zones of atoms that cause the strong repulsive barriers to the approach of one nucleus to another (like trying to force the same poles of a magnet together) it is the positive protons concentrated in the much smaller nucleus.  This force is called the Coulomb force and is very strong down to as little as $10^{-16}$ meters[132] which is about 1/10$^{th}$ of the diameter of an atomic nucleus.  The Coulomb force is the basic reason that chemistry exists.  We have a periodic table of elements, integrated, recognizable building blocks of the more complex molecules that make up the more complex matter in the universe.  The nuclei don't fuse but form chemical bonds, or what one may consider as intermediate stable states or entities that are again, unique and recognizable.

There is yet another barrier to overcome as one nucleus intimately approaches the other and that is the Strong Nuclear Force[133].  Many non-scientists have not been forced to consider the four basic forces in the universe.  Listed in order of strength they are:

- the Strong Nuclear Force
- the Weak Nuclear Force
- Electromagnetism (basically Coulomb force)
- and finally, the weakest force, Gravity.

In grade-school all of us learned about electromagnetism and gravity but it seems that the very good stuff, the nuclear forces were kept from us, perhaps as a later incentive to learn more[134] or perhaps as punishment for not being inquisitive enough or insistent that our education was more encompassing and complete.

The Strong Nuclear Force is the force that keeps nuclei together but it only works at very close distances.  The Strong Force buffers the Coulomb repulsion expected for protons and permits the formation of a

---

[132] I have deliberately used more accessible, but generally reliable, sources for scientific references in order to make reading and comprehension, easier.  See Coulomb Force in
http://www.britannica.com/EBchecked/topic/140084/Coulomb-force.

[133] http://aether.lbl.gov/elements/stellar/strong/strong.html.

[134] I make no apology for my deliberate mocking of most early science education … usually not administered by scientists or people who respect the native intelligence of their audience.

stable nucleus but this buffer must be overcome in order to unite two nuclei, so a large amount of energy is required to surmount the barrier.

When the two nuclei finally get close enough together the entire repulsive resistance is gone and there is a very large surplus of energy caused by the loss of a small amount of mass.

As shown in the diagram[135] for two hydrogen atoms (p-p interaction) there is a very considerable barrier that must be overcome with a significant amount of energy as they move together. The y-axis is potential energy and the x-axis is distance in femtometres[136]. The charge radius of a proton is approximately 0.84–0.87 femtometres. That is exactly where there is a precipitous drop in repulsion.

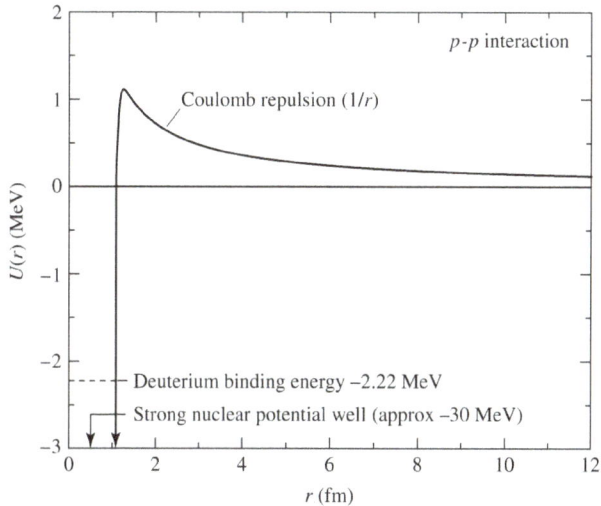

It is much easier for a neutron to penetrate a nucleus and fuse because there is essentially no Coulomb repulsion since it is neutral. That is why neutrons are the prime activators in nuclear fission because they can reasonably easily enter a nucleus and cause the split. The problem is very much greater for fusion because of the energy barrier to bringing two protons (hydrogen nuclei) together.

It is necessary to inject huge amounts of heat and to accelerate the protons into each other. By necessity this creates **plasma**[137] – a highly energetic ionized gas with its electrons free of their nuclei. Regular solid materials cannot tolerate the energy of a plasma as they would simply be ionized. Magnetic containment is the most practical way to control a plasma. This is another practical difficulty

---

[135] http://burro.astr.cwru.edu/Academics/Astr221/StarPhys/coulomb.html.
[136] The femtometre (Danish: *femten*, "fifteen") is an SI unit of length equal to $10^{-15}$ meters.
[137] Plasma is the fourth state of matter. It is only recently that plasmas are included in early science courses as a state of matter but this is very odd since plasmas seem to be the most common state of matter in the universe (all stars are basically plasmas).

in creating a fusion reactor. The plasma that has been heated to induce fusion will create a very large excess of energy (more plasma) that needs to be contained and controlled – in turn, costing a large amount of energy. Therein lays the *problem* – a sustainable, safely contained, break-even reaction (or better).

As an extension of the cold war the Soviet Union and the United States both began to look into fusion power[138]. The first approach to a continuous power reactor was the USSR's Tokamak[139], which used a toroidal magnetic field to contain the very hot plasma from the fusion. A different approach was the American Stellarator[140] invented in 1950 by Lyman Spitzer and built one year later at Princeton University (in 1961, after declassification, this became the Princeton Plasma Physics Laboratory a US Department of Energy Facility). These were very much cold war projects aimed at trying to control nuclear fusion.

The United Nations International Conference on the Peaceful Uses of Atomic Energy in Geneva was held in 1995 and this initiative effectively led to the declassification of previously secret programs and the slow but eventual international scientific collaboration that was essential to developing fusion as a power source.

In 1957, a very important organization was created by the UN.

> "The International Atomic Energy Agency (IAEA) is an international organization that seeks to promote the peaceful use of nuclear energy, and to inhibit its use for any military purpose, including nuclear weapons. The IAEA was established as an autonomous organization on 29 July 1957. Though established independently of the United Nations through its own international treaty, the IAEA Statute, [1] the IAEA reports to both the United Nations General Assembly and Security Council." Source: http://en.wikipedia.org/wiki/International_Atomic_Energy_Agency

The IAEA is important because it is deliberately above and outside of national political interests of any particular country[141] – strictly speaking, even the UN. It is also one of the most accurate sources of data both on nuclear power and nuclear weapons despite the nature of the voluntary supply of data. **Nuclear power is the only type of power that has an international agency associated with it.** We are

---

[138] The time-line for fusion available in Wikipedia is quite useful.
http://en.wikipedia.org/wiki/Timeline_of_nuclear_fusion.

[139] Tokamaks were developed in the 1950s by Soviet physicists Igor Tamm and Andrei Sakharov based on an early concept.

[140] The Stellarator was built in 1951 at Princeton University. http://en.wikipedia.org/wiki/Stellarator.

[141] A remarkable example of this independence comes from the prelude to the second Gulf War (March 20 – May 1, 2003) when, in a speech given in Cincinnati on October 7, 2002, George W. Bush insisted that Iraq (Saddam Hussein) had essentially weapons of mass destruction (WMDs). On March 7, 2003, IAEA Director General Mohamed El Baradei and UNMOVIC Executive Chairman Hans Blix briefed the UN Security Council on Iraq inspections. They found absolutely no evidence of WMD's and this was indeed the case after as what was, to many, an unnecessary and very expensive war, proved.

managing nuclear in a much more global and comprehensive way than we manage other sources of energy.

In searching the literature and internet for fusion information it became clear that the Tokamak toroidal plasma containment concept was initially much more popular and promising than the Stellarator. There is an impressive list of Tokamak based reactors in operation today[142] and dating back to the first operational one, TM1-MH, in Prague in the 1960's.

Iter[143] is the largest Tokamak reactor ever planned. It is based in a very small town, Saint Paul-lez-Durance, in southern France. The collaborating nations are China, the European Union, India, Japan, Korea, Russia, and the United States. The plan is for a 35-year collaboration (in the author's opinion this an excessive time).

Despite the international collaboration on Iter, it now seems that there have been a number of inherent problems with the Tokamak containment system and there has been significant renewed interest in the Stellarator approach. There are several projects such as the Wendelstein 7-X being built in Greifswald, Germany by the Max-Planck-Institut für Plasmaphysik (IPP)[144], which was completed in 2015. There are other Stellarator experiments going on internationally but these are not at the level of progress shown at Wendelstein. The Helically Symmetric eXperiment – HSX[145], is a modular coil stellarator located in the Electrical and Computer Engineering department at the University of Wisconsin-Madison. It is quite small so the scope of work has to be appropriately focused.

There is also a long standing (since about 1988) fusion experiment called the Large Helical Device (LHD) that is run by the National Institute for Fusion Science[146] in Toki Japan. It is the second largest superconducting stellarator in the world. It uses a heliotron magnetic field originally developed in Japan.

---

[142] http://en.wikipedia.org/wiki/Tokamak.
[143] https://www.iter.org/. Retrieved 02.21.2017.
[144] https://www.ipp.mpg.de/w7x.
[145] The Helically Symmetric eXperiment – HSX. http://www.hsx.wisc.edu/. Retrieved 02/17/2017.
[146] Large Helical Device. Toki Japan. http://www.lhd.nifs.ac.jp/en/. Retrieved 02/21/2017.

# PERCEPTIONS - BOMBS VERSUS POWER

*"We cannot solve our problems with the same thinking we used when we created them."*

- Albert Einstein, German Physicist (March 14, 1879 - April 18, 1955)

It is important to point out that "nuclear chain reactions" are not only confined to stars and intentionally created weapons but **have occurred spontaneously / naturally on earth**. In 1956, Paul Kazuo Kuroda[147] suggested that a natural fission reactor may have once existed. One basis of the hypothesis was that the natural abundance of uranium may sometimes combine with local geological circumstances to produce a chain reaction if conditions were appropriate.

In 1972, after looking at samples from a uranium mine in Oklo, Gabon, a French physicist, Francis Perrin, found that the isotope ratios indicated that a natural chain reaction had occurred. On September 25, 1972, the French Commissariat à l'énergie atomique (CEA) announced that self-sustaining nuclear chain reactions had occurred about 2 billion years ago[148]. Since that time several other specific sites have been identified and all are at Oklo. **In the modern rhetoric of anti-nuclear activism this very important bit of earth geological history seems to have been lost**. Man could not have much of a part because it occurred about 2 billion years before we evolved! The essential point is that both fusion and fission are "natural". Our discussions then should be shaped by this reality not rhetoric about what is good or bad. "Nature" doesn't care!

Although history is not a fundamental discipline, like mathematics or physics, it would be foolish to forget the history behind the discoveries and evolution of the nuclear age. In the previous section we looked at fission and fusion and the development of both in close temporal association with war.

When asking the seemingly simple question **"What came first – nuclear power or nuclear weapons?"**, the answer may seem obvious, given the Oklo reactor. However, within the modern era the answer is, unfortunately, not as predictable or straightforward as one might initially believe.

To better understand the question, we must look back in time, so when would we start the history clock? Again, there are many potential answers but probably the best and simplest place to start would be the almost universally known equation:

---

[147] http://www.encyclopediaofarkansas.net/encyclopedia/entry-detail.aspx?entryID=6619.
[148] Alex P. Meshik, "The Workings of an Ancient Nuclear Reactor". *Scientific American*, January 26, 2009.

$E = mc^2$ (E = energy, m = mass, c = speed of light)

Albert Einstein derived mass–energy equivalence within the context of special relativity. It was first mentioned in 1905 in the fourth of his *Annus Mirabilis* ("extraordinary year") papers:

"Ist die Trägheit eines Körpers von seinem Energieinhalt abhängig?"

**["Does the inertia of an object depend upon its energy content?"]**

*Annalen der Physik* **18** (13): 639–643 (1905).

This starting time for our clock emphasizes the significance of the mass-energy equivalence. Small amounts of mass can ... under the right conditions ... yield enormous amounts of energy. To put this into perspective yet again, 1 gram (1/454 of a pound) of matter converted completely into energy would produce $9 \times 10^{13}$ Joules. To translate this a little further, we could convert the energy into power. Since watts = joules/sec, if the conversion was done in one second, it would be $9 \times 10^{13}$ watts = 90,000,000,000,000 = 90 gigawatts (GW) which makes the installed capacity of the Hoover Dam (2.08 GW) look very small.

It was theoretically clear that there was inherently enormous power available from the conversion of matter to energy. The interest of Einstein and other scientists was to understand the nature of matter, light and the universe especially as it began to unfold in the age of new physics and cosmology. **There was no martial interest from the onset of any of these ventures**.

The four papers of Einstein's "extraordinary year" (1905) are the cornerstone of the new age of physics:

1. The Photoelectric Effect.
2. Brownian Motion
3. Special Relativity
4. Mass – Energy Equivalence

Chemistry, the other core discipline, was a bit further behind – basically on a different clock. Chemistry got a good jump-start with Dmitri Mendeleev's publication of his Periodic Table of elements in 1869. As valuable a tool as his reorganization was, it was basically a descriptive work. Although it was very systematic in its comparison of properties and mixing ratios it did not represent a fundamental understanding of the atom necessary to build a real picture of nature. It did indeed show relationships of reactivity that were later related to electronic configuration. Mendeleev left blanks in many places and was not the first chemist to do so, but he was the first to use the trends in his table to predict the properties of missing elements including mass estimates.

Four lighter elements that he predicted[149] were:
1. ekaboron (Eb) ... scandium (Scandium oxide was isolated in late 1879 by Lars Fredrick Nilson),
2. ekaaluminium (Ea) ... gallium (discovered in 1875 by Paul Emile Lecoq de Boisbaudran),
3. ekamanganese (Em) ... technetium (isolated by Carlo Perrier and Emilio Segrè in 1937),
4. ekasilicon (Es) ... germanium (Clemens Winkler, 1886).

The prefix "eka" (Sanskrit for 1) signified one row down in the same family or group. Regardless of the fact that this was not a fundamentally based contribution it was even the more remarkable given that it was based on keen observation and possibly a high degree of faith in the original atomic theory of John Dalton (1803).

J.J. Thompson's discovery of the electron was generally known in October of 1897, less than 8 years before Einstein's four papers. However, the most basic mysteries of the atom would still take considerably more time to uncover. It was not until 1911 that Ernest Rutherford discovered the "proton" (really the positive nucleus of an atom). In essence, even six years after Einstein's miraculous year, there is still no real model of the atom!

The pieces of the puzzle were beginning to become clearer but it took considerable time for the new faces of physics, chemistry and astronomy to articulate themselves and to begin to inform and reshape each other.

Edwin Hubble was the first to prove that Andromeda was an independent galaxy. He did this by looking at Cepheid variable starts in Andromeda with the new 100-inch Hooker telescope on Mount Wilson, CA. The findings were first announced in the *New York Times* on November 23, 1924. Very shortly afterwards Hubble found that the further away galaxies were from us the faster they were moving away. There was no other interpretation than the universe was expanding. This, in turn suggested "a beginning" hence, Big Bang. The newly expanding universe was of major interest to Einstein and many other scientists. After Hubble's discovery that the universe was expanding, Einstein called his faulty assumption that the Universe is static his "biggest mistake"[150].

The neutron was only discovered in 1932 by James Chadwick[151]. Finally, a rational and self-consistent model for the atom could be created and tested. Without the neutron, there is no modern chemistry possible - certainly no nuclear chemistry. In fact, the neutron makes sense of all of the other observations of matter and particularly of nuclear interactions.

---

[149] http://en.wikipedia.org/wiki/Mendeleev%27s_predicted_elements.

[150] Donald Goldsmith, "Einstein's Greatest Blunder? The Cosmological Constant and Other Fudge Factors in the Physics of the Universe", Harvard University Press: Cambridge, Mass., 1997.

[151] What is truly outstanding is that James Chadwick had been a student of Ernest Rutherford who had been a student of J.J. Thompson. Basically, the atom was "discovered" at the University of Manchester which was an absolute engine of physics at the beginning of the 20th century.

Very quickly after the discovery of the neutron it was realized that neutrons could initiate fission and possibly cause chain reactions.

This seemingly haphazard progress is typical of science and how we come to a higher level of understanding of natural phenomena. Clearly this is not a linear or comprehensive temporal account but it is highly indicative of reality. **Progress and understanding are not linear.** Even in this totally superficial sweep it seems that there was initially only legitimate scientific questioning involved in trying to sort out the atom, mass, and energy. In attempting to discuss nuclear issues or anything complex, for that matter, there are two essential things to keep in mind: **history is not linear and people generally do not behave rationally but emotionally.**

Acceptance of this statement is fundamental to understanding the following tale of the father of the Nobel Prize. Alfred Nobel (October 21, 1833 – December 10, 1896) was born in Stockholm, Sweden. He was a chemist, engineer, and inventor with a large number of patents. One of his several hundred inventions was dynamite (1867), which is perhaps the most famous and was likely the most financially successful.

In 1894, Nobel began his venture as a major arms manufacturer by taking over the Bofors iron and steel mill. Alfred Nobel had a particular talent and skill in designing munitions and has been credited with finding ways to stabilize nitroglycerine, inventing a detonator (1863), blasting caps (1865), gelignite (1875) and a number of other munitions. However, he clearly had other skills and sensibilities. These became apparent later in his life. In the 1870's, he became a good friend of Baroness Bertha Sophie Felicita von Suttner (née Countess Kinsky von Chinic und Tettau born on 9 June 1843, Prague, Austrian Empire). She was a particularly important personality as she wrote the very influential anti-war novel entitled "*Die Waffen nieder!*" ("*Lay Down Your Arms!*", 1889). She and Alfred Noble corresponded for years on the subject of peace.

Another telling event in Nobel's life occurred in 1888 when he read his own obituary in a French Newspaper that stated "Le marchand de la mort est mort" ("The merchant of death is dead")[152]. It was actually the death of Alfred's brother Ludvig in Cannes that was mistakenly reported as his. The article is said to have disconcerted Nobel and made him pensive about how he would be remembered. This supposedly led him to change his will (but he had been known to change his will often so there is perhaps some dramatic overstatement involved).

---

[152] 12th April 1888 - the premature obituary was published in the Paris newspaper "Ideotie Quotidienne" ("daily nonsense") under the headline "Le Marchand de la Mort est Mort" ("drummer of death is dead" and/or "the merchant of death is dead").

What is clear is that he was keenly aware of the growing peace movement in Europe and was personally linked to Baroness Suttner who later became known as the "generalissimo of the peace movement". In 1891, she established the Austrian Peace Society.

The Nobel Institute contends that:

> "There is little doubt that von Suttner's friendship with Alfred Nobel had an impact on the contents of his will, and many give her the credit for his establishment of a peace prize. "Inform me, convince me, and then I will do something great for the movement", Alfred Nobel said to Bertha von Suttner."[153]

She was the first woman to receive the Nobel Peace Prize in 1905. She died on June 21, 1914, in Vienna, Austria, still adamantly against the war that had just begun.

On 10 December 1896, at the age of 63, Alfred Nobel died in his villa in San Remo, Italy, from a cerebral hemorrhage. He left 94% of his assets to establish the five original prizes.

Nobel was sophisticated and doubtless understood the cut and thrust of European politics that had helped him make his fortune. However, it is far too simplistic to cast the man in one image. It is somewhat demeaning and trivial to suggest that his reaction to a mistaken obituary would motivate a sudden epiphany. It is more likely that the reasons are many and far more complex.

**This small aside represents the difficulty of dissecting most serious issues in that the motivations and actions of all people are often very difficult to de-convolute.** Was Nobel worried about people's perceptions of him or his fundamental tenets? He was intelligent and practical as most inventors are. Does dynamite kill people? Yes! Does dynamite improve the safety and efficiency of mining and construction? Yes! Are these mutually exclusive purposes? Not necessarily.

It is important to understand, or at least to probe, the difference between the intention to make bombs or to use the power available for other purposes. There is another problem with the utter simplicity of this approach in that it does not factor in the use of bombs or explosives for non-combative purposes. Dynamite has done just as much or more for mining than it has for death.

---

[153] http://www.nobelprize.org/nobel_prizes/peace/laureates/1905/suttner-facts.html. This reference was abstracted online 10/19/2013.

**Bombs versus Power?** That is the title of the chapter but, more importantly, it is the introduction to a very convoluted history, often both emotional and rationally compelling. Can we de-convolute the two streams of thought? Is it possible to be rational about nuclear power? *This is a critical question for our future. I would contend that it is essential to think incredibly clearly about these questions.*

Perhaps it is useful to take another short topical aside offered up by the NRA (National Rifle Association) "Guns don't kill ... people kill people". Although most people take this as an official slogan of the NRA it is not. However, it has been very effectively used since the late 1950's or early 60's.[154]

In his HBO comedy special, *Dress to Kill* (1999), Eddie Izzard added an interesting twist to the logic by saying, "The National Rifle Association says that guns don't kill people, people do. But I think the gun helps, you know? I think it helps. I think just standing there going 'BANG!' — that's not going to kill too many people, is it?"

These opposing points of view are a major problem in the discussion of nuclear power since it has been highly associated with nuclear weapons. **Conflation and polarization are powerful social and political tools**. As long as one keeps muddying the waters, resolution of fundamental problems or conflicts becomes more difficult or sometimes impossible.

No inanimate objects kill people without external forces. In general, the major external force is other people but it is somewhat naïve to think that hiding all weapons will stop intra-species violence it will just shift the statistics. Knives are not as effective as guns that enable a more rapid and remote delivery of deadly force.

In order to intelligently discuss the issues, the history and context of nuclear development must be clearly understood.

Going back to the inception of the Nobel Prizes in Alfred Nobel's will of 1895:

> "The whole of my remaining realizable estate shall be dealt with in the following way: the capital, invested in safe securities by my executors, shall constitute a fund, the interest on which shall be annually distributed in the form of prizes to those who, during the preceding year, shall

---

[154] May 31, 1959 Associated Press story found in NewspaperArchive.com, Fred A. Roff Jr., president of the Colt Patent Fire Arms Co., was quoted as saying "Our big concern is to make sure that guns get into the hands of only those who know how to use them. Guns don't kill people. People kill people."

have conferred the greatest benefit on mankind. The said interest shall be divided into five equal parts, which shall be apportioned as follows: one part to the person who shall have made the most important discovery or invention within the field of physics; one part to the person who shall have made the most important chemical discovery or improvement; one part to the person who shall have made the most important discovery within the domain of physiology or medicine; one part to the person who shall have produced in the field of literature the most outstanding work in an ideal direction; and one part to the person who shall have done the most or the best work for fraternity between nations, for the abolition or reduction of standing armies and for the holding and promotion of peace congresses."

Physics, Chemistry, Physiology & Medicine, Literature and Peace are the original prizes[155] to be given "to those who, during the preceding year, shall have conferred the greatest benefit on mankind**. If Nobel, the father of dynamite, could voice so clearly and admirably humanitarian goals we should take heed.**

Another related and perplexing question is, do Nobel Peace Prizes contribute to world peace? This is not a clean or elegant segue into what I would like to discuss and that is - the general effects of the aftermath of WWII – the "peace", the Cold War and nuclear proliferation.

Before the cold war began to heat up, the creation of two major extended international conflicts was already determined by the *de facto* collapse of the British Empire. The first issue was that of Palestine. The formation of the state of Israel on May 14, 1948 was announced by David Ben-Gurion one day before the official end of the British Mandate for Palestine[156] was to terminate. The problem with Israel and the Mandate for Palestine began after the fall of the Ottoman Empire after WWI[157]. Within one day of the declaration of Israel as a country the first Arab-Israeli War[158] was precipitated by the invasion by three Arab nations (Egypt, Jordan and Syria, together with expeditionary forces from Iraq). This immediate and persistent threat was a clear indication that Israel was prepared to go to great lengths to defend itself and that is still very clear today.

---

[155] It is a personal peeve that "Economics" is a prize category. Economics would have barely qualified as a discipline to Nobel and, as a field, could not ever fulfill the intentions of Nobel in "conferring the greatest benefit on mankind".

[156] "The roots of separatism in Palestine: British economic policy, 1920-1929", Barbara Jean Smith, Syracuse University Press, 1993.

[157] The British Mandate for Palestine, was a legal commission for the administration of the territory that had formerly constituted several areas of the Ottoman Empire. The draft of the Mandate was formally confirmed by The Council of the League of Nations on July 24, 1922, and later supplemented with the Transjordan memorandum that came into effect on September 29, 1923 following the ratification of the Treaty of Lausanne.

[158] http://en.wikipedia.org/wiki/1948_Arab%E2%80%93Israeli_War.

The Arms Control Association states, on its website[159], that:

> "Israel has not publicly conducted a nuclear test, does not admit to or deny having nuclear weapons, and states that it will not be the first to introduce nuclear weapons in the Middle East. Nevertheless, Israel is universally believed to possess nuclear arms, although it is unclear how many weapons Israel possesses."

The motivation for Israel's defensive posture is clear in the continued conflict in the middle-east since the birth of the state of Israel on May 14, 1948. The local tensions have been increasing as it has become apparent that Iran is a nuclear power or certainly intends to be one regardless of ongoing international attempts at curbing those aspirations. In Israel, this is taken as a credible threat given prior and ongoing Iranian support for Hamas[160] and its' continued conflict with Israel.

The second significant event was the successful independence movement in India that was symbolized by Mahatma Gandhi and Jawaharlal Nehru, who became the first prime minister of India (August 15, 1947 – May 27, 1964)[161]. With widespread strikes and non-violent protests, the movement made it effectively impossible for the British, who were much weakened by WWII, to hold onto India. The Indian Independence Act was passed in the Parliament of the United Kingdom and received royal assent on July 18, 1947. The new countries of Pakistan (for Muslims) came into being on August 14, and India (for Hindu's) on August 15, 1947[162]. On the surface, the act seemed to be a good compromise but the devil was in the details and partitioning of some areas such as Kashmir and Bengal were problematic then and still are the centers of conflict today.

Besides the enmity between Pakistan and India there were other regional forces and events that drove India to seek nuclear weapons. The other primary reason for Indian insecurity is China. India lost territory to China in a brief Himalayan border war in October 1962 and that was certainly an impetus for pushing weapons testing within their 3-part nuclear program. This fear of the Chinese is widespread in India at present[163].

---

[159] http://www.armscontrol.org/factsheets/Nuclearweaponswhohaswhat.

[160] http://www.telegraph.co.uk/news/worldnews/middleeast/iran/11515603/Iran-is-intensifying-efforts-to-support-Hamas-in-Gaza.html. Article of Thurs. 28 May 2015.

[161] In June of 1946 (before becoming Prime Minister) Nehru said, "As long as the world is constituted as it is, every country will have to devise and use the latest devices for its protection. I have no doubt India will develop her scientific researches and I hope Indian scientists will use the atomic force for constructive purposes. But if India is threatened, she will inevitably try to defend herself by all means at her disposal." Source: B. M. Udgaonkar, *India's nuclear capability, her security concerns and the recent tests*, Indian Academy of Sciences, January 1999.

[162] Hoshiar Singh, Pankaj Singh; Singh Hoshiar. *Indian Administration*. Pearson Education India. p. 10. ISBN 978-81-317-6119-9.

[163] "72% of Indians fear border issue can spark China war", *Times of India*, July 15, 2014. http://timesofindia.indiatimes.com/india/72-of-Indians-fear-border-issue-can-spark-China-war/articleshow/38397343.cms.

The consequence of Indian nuclear testing was the inevitable provocation for Pakistan to do the same.

Another very interesting potential contributor to the tensions is the symbolic value of elevating Indian status to the level of a potentially serious player at the international poker table. This is not to be discounted out-of-hand when officials within the nuclear program voiced distinct pride in seeing India no longer patronized by the major powers. There are some good insights into the Indian nuclear program in *India's Nuclear Bomb: The Impact on Global Proliferation* by George Perkovich (University of California Press, 2001) but one must take into account that this is not a technical but historical approach.

The contention of the former superpowers, the Soviet Union and particularly the United States, that proliferation of nuclear weapons can be controlled by economic sanctions and world opinion is specious and would be quite laughable were the situation not so tense and fraught with potential dangers. The implicit contention is that the powers with the nuclear weapons would only use them in the direst of circumstances. There are many who might argue that the only two military uses of nuclear weapons were justified at Hiroshima and Nagasaki[164]. That horrible power and death toll were burned onto the consciousness of the world. However, analytical minds must be more encompassing in their analyses to understand the present world as it stands.

The Treaty on the Non-Proliferation of Nuclear Weapons (aka Non-Proliferation Treaty or NPT) is an international treaty that was opened for signature in July of 1968, and was brought into force in 1970. The treaty was extended indefinitely on May 11, 1995. Although 191 states have joined the NPT and it is arguably one of the most successful efforts of its type in history, the non-members are the real problem. North Korea withdrew in 2003 and India, Israel, Pakistan and South Sudan never joined[165]. The situations for Israel, India and Pakistan are somewhat understandable. It would be absolutely incredible if the South Sudan could or would ever have a nuclear program given their own internal problems. It is North Korea that remains the largest uncertainty.

Upon looking back at the Korean War (June 25, 1950 – July 27, 1953) and the US involvement there it must be emphasized that there was a United Nations military intervention on behalf of South Korea whereas North Korea was supported by the Chinese and the Soviet Union. Although this may be considered a proxy war and an extension of the cold war, the origins and regional history are much more complicated. The result of the Korean War was a divided Korea and a more-or-less entrenched quasi-wartime situation. As the economic success of South Korea grew the poverty of the north kept pace. It was clearly not in the interests of either China, South Korea or the United States to make the

---

[164] "Little Boy" was dropped on Hiroshima on August 6, 1945. "Fat Man" was dropped on Nagasaki on August 9, 1945.
[165] http://en.wikipedia.org/wiki/Treaty_on_the_Non-Proliferation_of_Nuclear_Weapons.

Korean peninsula a nuclear point of contention. The nuclear weapons program of North Korea is very much more about the political system within the country and, at least in part, due to the bizarre "cult of the emperor"[166] set up Kim Il-sung (president from 1972 until his death, July 8, 1994), continued by his son Kim Jong-il (until his death on December 17, 2011), and his grandson Kim Jong-un (to the present). One North Korean motivation is to be considered more seriously by the world's major powers and another seems to be relatively simplistic sabre rattling.

The Soviet Union was asked by North Korea (in 1962) for help to develop nuclear weapons and was refused but did agree to help with a civilian nuclear power program. China also refused to help North Korea develop weapons. These actions cannot be cast, in any manner, as an extension of the cold war. It seems apparent that both the Soviet Union and China had their separate reservations about the North Korean regime. It was also not in China's interest to destabilize the Korean Peninsula any further.

The cold war certainly fueled the general tension of the public and the threat of global nuclear war. In 1951, the Federal Civil Defense Administration produced a particularly chilling and telling video entitled *Duck and Cover*[167, 168] that featured an animated "Bert the Turtle" who would retreat into his shell with the threat of danger. The video was produced in association with the Safety Commission of the National Education Association. The video was seen by millions of school children across the United States and likely many in Canada and the United Kingdom. It is not surprising that intrinsic fear was seeded within an entire generation. Apparently, this public reaction was in response to active Soviet nuclear tests. I remember having grown up with this propaganda, stating, much to my parents and teacher's despair, "duck and kiss your ass goodbye". Being a scientist with a nuclear background I can say with some authority the statement that I had uttered as a child was definitively correct if not somewhat inappropriate in a school room.

At the end of WWII only the USA had atomic bombs. However, due to the Manhattan Project, with its associated British and Canadian scientists there was already a broadening of the nuclear weapons information base. To this mix must also be added the scientists of the Soviet Union who were very aware of Nazi nuclear initiatives. There was, in fact, a race by both sides as WWII ended to secure as many Nazi scientists as possible. It was inevitable that the Soviets would feel the absolute necessity of arming themselves with equivalent weapons. There are really many more articulated political and economic reasons the Soviet Union felt that it needed parity with the United States.

Certainly, a major portion of the blame for the arms race and perhaps even the Cold War itself may be laid at the feet of the United States. The Soviets were convinced that the US initiatives to keep a large

---

[166] The "cult of the emperor" was prevalent in ancient Rome and was an active acceptance of the imperial family as deities.

[167] See the YouTube video https://www.youtube.com/watch?v=IKqXu-5jw60.

[168] The film production details are available at http://en.wikipedia.org/wiki/Duck_and_Cover_%28film%29.

peacetime army, build naval bases around the world, control oil resources in the middle east and a number of other measures, were positive proof that the US was intent on world domination[169]. The "Truman Doctrine" (as outlined by the Soviets) was a potential problem for many countries in the world aside from the Soviet Union.

The disturbing concept of a "strategic balance of terror" was introduced in 1956 in a book by Glenn Snyder titled *The Balance of Power and the Balance of Terror*[170]. In it, the basic premise is that offsetting nuclear arsenals reduce the threat of nuclear war but increase subsidiary or less severe conflicts. This became known as the **stability-instability paradox** that has been discussed extremely widely since that time. There seems to be some quite convincing quantitative data, presented in a 2009 paper by Robert Rauchhaus[171], that the basic paradox is not simply an international relations theory but is indeed true.

Another somewhat ironic twist to the stability-instability paradox is that nuclear weapons can be seen as both good and bad. The threat of mutual destruction is a powerful incentive not to use the nuclear weapons but it creates many difficulties in the form of proxy wars and secondary confrontations. In addition, it has made it difficult for any of the players to step down and to attempt to change the nominal and potentially fragile *status quo*. The India-Pakistan situation falls into this paradox[172].

Even with these reservations and periodic setbacks there has been progress at stabilization of the largest nuclear arsenals. The Cold War effectively lasted from 1947 to the dissolution of the Soviet Union on December 26, 1991. There was certainly a thawing of Cold War relations as the superpowers began to realize the costs and ridiculous scale of the Cold War. There were two rounds of Strategic Arms Limitation Talks (SALT I and SALT II). On November 17, 1969, the formal SALT I talks began in Helsinki, Finland. It took over two and a half years to work out the details of the proposed limitations and a verification scheme. President Richard Nixon and Soviet General Secretary Leonid Brezhnev signed the ABM Treaty and interim SALT agreement on May 26, 1972, in Moscow. SALT II talks began late in 1972 but there were still some unresolved issues from SALT I and further political complications on both sides. Finally, On June 17, 1979, President Carter and Brezhnev signed the SALT II Treaty in Vienna. However, there was widespread congressional skepticism of the treaty and particularly Soviet intentions and congress was reluctant to ratify the treaty. The reluctance became refusal when the Soviet Union invaded Afghanistan on January 3, 1980.

---

[169] "Soviet Ambassador Nikolai Novikov on the U.S. Drive for World Supremacy, September 1946". In **Major Problems in American History Since 1945**. Edited by Robert Griffith and Paula Baker. Wadsworth, Boston, 2007. p. 43-46.

[170] Snyder, Glenn (1965). *The Balance of Power and the Balance of Terror*. Chandler.

[171] Rauchhaus, Robert (2009). "Evaluating the Nuclear Peace Hypothesis A Quantitative Approach". *Journal of Conflict Resolution* **53** (2): 258–277.

[172] Narang, Vipin. "Military Modernization and Technological Maturation, an Indian Perspective: Stabilizing the Instability-Stability Paradox". *Nuclear Learning in South Asia: The Next Decade*. Naval Postgraduate School, June 2014. p. 48-57.

After a number of pages of what might seem like rambling discourse are we any less confused or more informed than we started? What did come first, bombs or power? Is the question really bombs versus power? Will our futures consist of needless focus on perhaps irrelevant and/or unresolvable questions? My personal preferences are to engage in a discussion of solutions.

The "cat is out of the bag" and it is very difficult to herd cats or to get them back into the bag. It is most intelligent peoples' wish to focus on solvable problems. My contention is that although nuclear weapons are indeed intrinsically terrifying, our fear of nuclear war or the use of these weapons is irrationally high given the fact that only two have ever been used against an opponent and that was at the very beginning when there was a war in progress and no full or sober understanding of the consequences of using the weapons existed until they were used. The bombing of Hiroshima and Nagasaki shaped the 20th century in essential ways. Of, course there must be every legitimate effort to improve non-proliferation efforts and to reduce existing arsenals and some of those efforts have been successful. However, it is a new century[173] with an entirely new set of hazards and priorities.

By looking at this rather superficial account it is reasonably clear that **the "nuclear issue" is NOT an issue of BOMBS versus POWER**. The real and important issue is that we start to look at nuclear power in a rational way and take the discussion of nuclear weapons out of the picture as completely as possible. Of course, there are problems, like nuclear waste management, appropriate selection of reactor systems, better control of nuclear materials as well as many more but these are well within the realm of solvable problems.

*The major hazard and real "threat" of the new century is global warming. The development of nuclear power (especially fusion) is essential to a balanced consideration of options for climate stabilization and balancing of energy requirements – it is of the highest importance.*

---

[173] Forrest E. Morgan, Karl P. Mueller, Evan S. Medeiros, Kevin L. Pollpeter, Roger Cliff. *Dangerous Thresholds: Managing Escalation in the 21st Century*. Rand Corporation, July 29, 2008.

# FISSION

Perhaps one of the more minor consequences of fusion is fission. When heavy elements are created there are often several isotopes of an element that are formed. Some of the isotopes are not intrinsically stable and those nuclei will eventually break into fragments – fission. Any isotope that undergoes a nuclear fission when bombarded with neutrons is a "fissile"[174] isotope.

Fission reactions occur naturally and spontaneously but generally not dramatically. In fact, radioactive decay is one of the three main reasons for the internal temperature of the earth[175].

The most familiar and important fission reactions for power production and for weapons are undergone by $^{235}U$ (Uranium 235) and $^{239}P$ (plutonium-239). It is important to note that $^{235}U$ is naturally occurring and $^{239}P$ does not exist in any significant quantity in nature. To produce power the fission must be controlled or "moderated" but for weapons the intent is to release all of the energy immediately. To produce any significant energy from fission one must induce a "chain reaction" in which the fission of one nucleus induces the fission of at least one other nucleus. This chain reaction is propagated by the neutrons that are released in the initial fission.

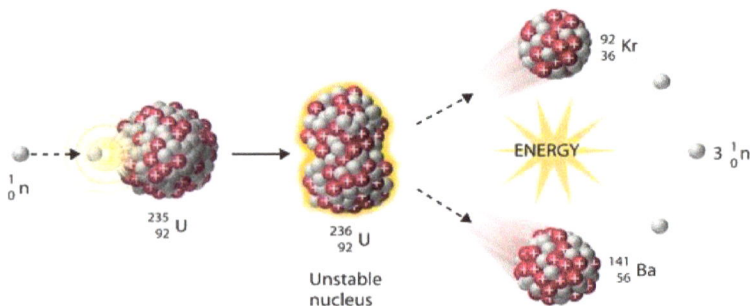

$^{1}_{0}n$    $^{235}_{92}U$    $^{236}_{92}U$ Unstable nucleus    ENERGY    $^{92}_{36}Kr$    $^{141}_{56}Ba$    $3\,^{1}_{0}n$

A TYPICAL $^{235}U$ FISSION (FROM FISSION FIG 20.7 CHEMWIKI.UCDAVIS.EDU)

---

[174] Most actinide isotopes with an odd neutron number are fissile. Most nuclear fuels have an odd atomic mass number (A = Z + N = the total number of nucleons), and an even atomic number (Z). For more on fissile materials see http://en.wikipedia.org/wiki/Fissile_material.

[175] (1) extraterrestrial impacts, (2) gravitational contraction of the Earth's interior, and (3) the radioactive decay of unstable isotopes are responsible for earth's internal temperature. These were all more important in the early stages of Earth's development but it is accepted that radioactive decay is currently the most important factor in maintaining heat since impacts and contraction have reduced disproportionately.

During nuclear fission, the nucleus usually divides asymmetrically. What is interesting is that every fission event of a given nuclide does not give the same products. There are more than fifty (50) different fission modes identified for $^{235}U$. Therefore, the fission of a nuclide cannot be described by a single equation and is much better described by a statistical distribution of many pairs of fission products. In general, the mass ratio of each pair of fission products produced by a single fission event is roughly 3:2[176].

Chicago Pile-1 (CP-1) was the world's first nuclear reactor constructed for the Manhattan Project[177] and was executed by the Metallurgical Laboratory at the University of Chicago. It was built under the west viewing stands of the original Stagg Field that was no longer in use. On December 2, 1942, Enrico Fermi supervised the first sustained nuclear chain reaction. Fermi described the apparatus as "a crude pile of black bricks and wooden timbers." It had no heat shield or cooling system and, in any era, would have been considered dangerous. The site is now a National Historic and Chicago Landmark.

The first truly functioning fission reactor was Chicago Pile-3 (CP-3) that went critical on May 15, 1944. It was the first heavy water reactor in the world and used natural uranium as the fuel.

Canada was not at all far behind and its start was at a joint British-Canadian laboratory, the Montreal Laboratory, in 1942. The National Research Council of Canada developed a design for a heavy water nuclear reactor. This reactor was called National Research Experimental and was the most powerful research reactor in the world when completed. It is difficult to determine but highly likely that the Canadian design had a role in shaping the CP-3 heavy water reactor. In 1944, approval was given to proceed with the construction of the small ZEEP (Zero Energy Experimental Pile) test reactor at Chalk River, Ontario and on September 5, 1945 at 3:45 p.m., the 10-Watt ZEEP successfully achieved the first self-sustained nuclear reaction outside the United States[178].

The intent here is not to present the alternative choices of reactors (that will be done in a later chapter) but to illustrate the very close temporal proximity of weapons production and the construction of the fission reactors.

---

[176] A more complete description is available from a number of sources but a particularly clear one is http://chemwiki.ucdavis.edu/Physical_Chemistry/Nuclear_Chemistry/Nuclear_Reactions.

[177] There are some interesting details of the first run of CP-1 at https://www.osti.gov/opennet/manhattan-project-history/Events/1942-1944_pu/cp-1_critical.htm.

[178] Canada Science and Technology Museum, *ZEEP — Canada's First Nuclear Reactor*, Ottawa: Canada Science and Technology Museum.

In the world, nuclear power is being used safely and effectively. A much more essential point is that there are many countries in the world with nuclear power reactors and no intention of making, having or using nuclear weapons.

Perhaps the renewed interest, ability to use effective superconductors in the plasma containment, and the urgency of dealing with the global overuse of fossil fuel burning will bring about a viable and robust fusion reactor design. It has been about 63 years since the first H-bomb detonation so it would be a very welcome step forward to have a functioning fusion power reactor in the near future, rather than just a new series of experiments.

Fission is just a little bit like fusion in that it is much easier to make a bomb, basically an uncontrolled reaction, than it is to create power in a controlled and safe manner. That is where the similarities end because making a safe nuclear reactor using fission is a relatively easy task compared to making a fusion reactor. The essential words here are **controlled** and **safe**.

The list of those countries with totally peaceful nuclear policies are: Canada, Japan, Sweden, Germany, Belgium, Spain, Switzerland, Finland, Hungary, Slovakia, Brazil, Bulgaria, Mexico, Romania, South Africa, Armenia, Netherlands and Slovenia. **The majority of countries with nuclear power are committed to peace and a cleaner planet**. Sweden is the country with highest percentage of power from nuclear reactors calculated on a *per capita* basis. This can be seen in the table below that was constructed based on the IAEA reactor list.

## IAEA PRIS – Operational Nuclear Power Reactors

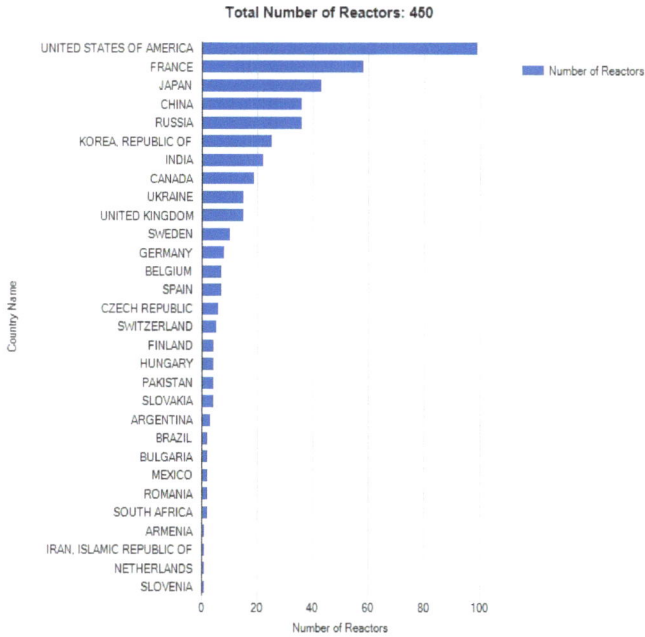

Total Number of Reactors: 450

https://www.iaea.org/PRIS/WorldStatistics/OperationalReactorsByCountry.aspx. Retrieved 11/16/2016.

NOTE: This graphic does not contain the 6 reactors for Taiwan that are listed by the IAEA but separately. Clearly the children are playing politics at every level! The reactors are included in the data table below.

## IAEA PRIS - Operational Nuclear Power Reactors
### Ranked by nuclear power produced based on population

| Country | Reactors | Total Net Electrical Capacity [MW] | 2015 Population (thousand) | MW / 1K Persons |
|---|---|---|---|---|
| SWEDEN | 10 | 9651 | 9,798.87 | 0.98 |
| FRANCE | 58 | 63130 | 66,808.38 | 0.94 |
| BELGIUM | 7 | 5913 | 11,285.72 | 0.52 |
| FINLAND | 4 | 2752 | 5,482.01 | 0.50 |
| KOREA, REPUBLIC OF | 25 | 23133 | 50,617.04 | 0.46 |
| SWITZERLAND | 5 | 3333 | 8,286.98 | 0.40 |
| CANADA | 19 | 13524 | 35,851.77 | 0.38 |
| CZECH REPUBLIC | 6 | 3930 | 10,551.22 | 0.37 |
| SLOVAKIA | 4 | 1814 | 5,424.05 | 0.33 |
| SLOVENIA | 1 | 688 | 2,063.77 | 0.33 |
| JAPAN | 43 | 40290 | 126,958.47 | 0.32 |
| UNITED STATES OF AMERICA | 99 | 99868 | 321,418.82 | 0.31 |
| UKRAINE | 15 | 13107 | 45,198.20 | 0.29 |
| BULGARIA | 2 | 1926 | 7,177.99 | 0.27 |
| TAIWAN, CHINA | 6 | 5052 | 23,490.00 | 0.22 |
| HUNGARY | 4 | 1889 | 9,844.69 | 0.19 |
| RUSSIA | 36 | 26557 | 144,096.81 | 0.18 |
| SPAIN | 7 | 7121 | 46,418.27 | 0.15 |
| UNITED KINGDOM | 15 | 8918 | 65,138.23 | 0.14 |
| GERMANY | 8 | 10799 | 81,413.15 | 0.13 |
| ARMENIA | 1 | 375 | 3,017.71 | 0.12 |
| ROMANIA | 2 | 1300 | 19,832.39 | 0.07 |
| ARGENTINA | 3 | 1632 | 43,416.75 | 0.04 |
| SOUTH AFRICA | 2 | 1860 | 54,956.92 | 0.03 |
| NETHERLANDS | 1 | 482 | 16,936.52 | 0.03 |
| CHINA | 36 | 31402 | 1,371,220.00 | 0.02 |
| IRAN, ISLAMIC REPUBLIC OF | 1 | 915 | 79,109.27 | 0.01 |
| MEXICO | 2 | 1440 | 127,017.22 | 0.01 |
| BRAZIL | 2 | 1884 | 207,847.53 | 0.01 |
| PAKISTAN | 4 | 1005 | 188,924.87 | 0.01 |
| INDIA | 22 | 6225 | 1,311,050.53 | 0.00 |
| **Total** | **450** | **391915** | | |

Columns A-C source:
https://www.iaea.org/PRIS/WorldStatistics/OperationalReactorsByCountry.aspx
Column D source: http://data.worldbank.org/indicator/SP.POP.TOTL
Column E source: Calculated by author
Population of Taiwan from http://www.taiwan.gov.tw/content2.php?p=29&c=48
Population data for Taiwan was absent from the World Bank Data set – likely for political reasons.

The revised listing of countries and reactors above provides more than a little food for thought. For example, it is quite interesting that a country such as Pakistan, with 4 power reactors produces so little actual power[179]. It is not surprising that Pakistan is not in the list of countries that have non-martial nuclear interest. Those countries that have opted for nuclear power without nuclear weapons have guaranteed the highest efficiency power production at the lowest environmental cost.

In summary:

- the world has many functioning and safe nuclear power reactors,
- very few countries with nuclear power have nuclear weapons or any interest in developing nuclear weapons,
- there are different reactor types developed for distinctly different reasons (to be discussed later). This segues into a discussion of clearly safer choices for power reactors.

---

[179] Having worked in the nuclear industry I can say with some certainty that this smells like a weapons program to me **OR** the reactors are very, very tiny or very, very poorly run.

# FISSION REACTOR SHOPPING LIST

> *"The odds of going to the store for a loaf of bread and coming out with only a loaf of bread are three billion to one."*
>
> Erma Bombeck, Author / Columnist (February 21, 1927 - April 22, 1996)

When one shops for anything it is very important to understand what the intended menu is! Are we looking for a supply of material for weapons or for a reliable power system or for both?

As a scientist who was brought up in the CANDU (Canadian Deuterium Uranium) reactor system I cannot claim impartiality. The CANDU concept is intrinsically the safest reactor system in the world – **the reactors cannot melt down**. The basic reason for this is that the fuel is non-enriched or natural uranium. This means that to maintain the critical neutron chain reaction the moderator or heat transfer agent must not absorb neutrons. Unlike the case of American pressurized light water reactors, a Canadian reactor uses heavy water. There's a very large difference in the two concepts in that, if the CANDU loses its moderator water the chain reaction will become subcritical – there will not be enough neutrons to continue a chain reaction. If a pressurized light water reactor loses coolant, there is the possibility of the China Syndrome also known as meltdown. At a minimum, a meltdown results in catastrophic damage to the reactor core elements and often results in release of radioactive material into the environment. At a maximum – KABOOM (probably a very dirty explosion at that). Fairly serious partial to complete meltdowns have been seen at Fukushima and Chernobyl whereas TMI (Three Mile Island) was notably less serious from both a damage point of view and radioactive releases. There is a specific discussion of nuclear accidents featuring the major ones in the world in the section describing the IAEA ratings.

## FUNDAMENTAL DIFFERENCES IN REACTORS

The most important first tier differences in reactor types are rooted in the selection of fuel and moderator and, in this sense, there are really two basic types:

| FUEL | MODERATOR | |
| --- | --- | --- |
| Natural Uranium | Heavy Water ($D_2O$) | ← *intrinsically safer* |
| Enriched Uranium | Light Water ($H_2O$) | |

The difference is simply between the very high neutron absorption of light or natural water and the almost opposite behavior of deuterium oxide or heavy water. The physics of this is quite simple. Heavy water is $D_2O$ and the deuterium is simply hydrogen that has taken on a single neutron. The heavy water is relatively difficult to convert to even heavier water - DTO or $T_2O$ (fully tritiated water). One must keep in mind that tritium (the third hydrogen isotope) is radioactive. It is not very "hot" or dangerous in terms of radiation because it has a half-life of 12.32 years[180] and its decay is by beta emission (ejection of an energetic electron). Tritium ($^3H$) loses one electron from one of the neutrons in the nucleus and this produces a proton and it jumps up one element in the periodic table to $^3He$ (helium 3, an isotope of helium with only one neutron instead of two).

Of, course there are many subtypes of reactors based on operational choices and it is disingenuous to suggest that the only basic selection criteria are "bombs or power" however, most of the other considerations are statistically unimportant or were already tied to another logic (i.e. socio-political imperatives that are often very local / national).

For example, research reactors, perhaps those used to create medically useful isotopes or for neutron activation analysis (NAA), start with the notion of needing relatively high neutron flux (number of neutrons per square centimeter per second) in order to produce isotopes from target materials. These small reactors are, in fact, so useful that there is an entire type called "Slowpoke" (Safe LOW-POwer Kritical Experiment), a low-energy, nuclear research reactor designed by Atomic Energy of Canada Limited (AECL) in the late 1960s. It has a very low critical mass but provides neutron fluxes higher than available from a small particle accelerator or other radioactive sources. It makes them very efficient for neutron activation analysis, in research, and as a commercial service, but also for teaching, training, irradiation studies, neutron radiography, and the production of radioactive tracers. They are reliable, easy to use, and have very predictable nuclear flux because the fuel remains the same for a minimum of 10 years. The energy spectrum of the neutrons produced is very reproducible (to about 1%). Only five of the original reactors are still in operation[181] and Canada seems to have lost the commercial market to the Chinese. These are small reactors, usually open pool, with highly enriched uranium fuel – not the formula for an easily maintainable power reactor.

Now, if we want a power plant for a nuclear submarine or aircraft carrier we are in good company in that all U.S. Navy submarines and supercarriers built since 1975 are nuclear-powered.[182] The first ship with a nuclear reactor was the submarine USS Nautilus (SSN-571). This represented a major step forward for marine propulsion and was particularly important to the military role of submarines. It is now very clear that the most potent nuclear threat / deterrent is from the "boomers" (nuclear missile carrying submarines)[183] that are difficult to detect, mobile and have MIRV'ed missiles[184]. Depending on

[180]Lucas, L. L. and Unterweger, M. P. (2000). "Comprehensive Review and Critical Evaluation of the Half-Life of Tritium". Journal of Research of the National Institute of Standards and Technology 105 (4): 541.

[181] https://en.wikipedia.org/wiki/SLOWPOKE_reactor. Retrieved 07/05/2015.

[182] https://en.wikipedia.org/wiki/United_States_naval_reactors. Retrieved 07/05/2015.

[183] The "boomers" of the US Navy presently officially number 18. They are all Ohio Class (SSBN 726) type submarine capable of carrying 24 missiles - Trident or Trident II (see footnote 132 on missiles). Source: http://www.navy.mil/navydata/cno/n87/today/ssbn.html retrieved 07/17/2015.

the particular missile being carried by US ballistic missile submarines, the maximum total number of possible warheads is 6048. The most realistic current estimate is 5472 warheads[185]. This number is more than enough to maintain "the strategic balance of terror". No land-based missiles are necessary.

Besides moving missiles around the world, the real benefit to shipping is the lack of need for fossil fuel, much less vessel mass per unit speed, virtually no pollution and no practical limit on the length of voyages, even submerged.

All of the Ohio Class "boomers" are powered by the S8G reactor (S = submarine, 8 = generation number of the core designed by the contractor, and G stands for General Electric one of several different contractors). The reactor is a pressurized water reactor (PWR)[186] with an extended fuel lifetime.

Propelling any large vessel such as an aircraft carrier would require tons of fossil fuel but all 10 of the Nimitz Class supercarriers are power by two A4W reactors[187] (each rated at 550 MW) with a projected core life of 20 years without refueling. These are considerable sized reactors. To get some idea of the actual power produced consider that the average domestic American home consumes about 2 KW of instantaneous power[188]. This means that 500 homes would consume only 1 MW. The reactors of a Nimitz Class aircraft carrier could power 1,100,000 homes … a good-sized city.

It would seem inconsistent that the US navy would risk the safety of its men and mission on potentially dangerous equipment. The nuclear navy is a reality because of the immense power and efficiency of using reactors instead of fossil fuels. Unfortunately, the technology has not been transferred to commercial shipping. It would be ironic if supertankers were nuclear powered! A bizarre concept but why not? Any real reduction in emissions of $CO_2$ is good for the planet. This could also be extended to all cruise ships – nasty smoky things!

---

[184] MIRV stands for **M**ultiple **I**ndependently [targeted] **R**e-entry **V**ehicles = many warheads. The submarine-based Trident missile generally has 8 warheads per missile and the Trident II has up to 14.

[185] There are currently 14 of the 18 SSBN subs using Trident II's. Source: https://en.wikipedia.org/wiki/UGM-133_Trident_II retrieved 07/17/2015.

[186] https://en.wikipedia.org/wiki/S8G_reactor. Retrieved 07/17/2015.

[187] https://en.wikipedia.org/wiki/A4W_reactor. Retrieved 07/17/2015.

[188] https://en.wikipedia.org/wiki/Domestic_energy_consumption. Retrieved 07/17/2015.

# INTERNATIONAL POWER REACTORS

It is quite instructive to examine the following table to see what countries are relatively dedicated to nuclear power.

**COUNTRIES RANKED BY % NUCLEAR ELECTRIC POWER**

| Country | Number of Reactors | Total Electricity GW.h (2014) | Nuclear Electricity GW.h (2014) | % Nuclear Electricity (2014) | Reactor Type AGR & GCR | Reactor Type BWR | Reactor Type FBR | Reactor Type HTGR | Reactor Type PHWR | Reactor Type PWR | Reactor Type OTHER* | Shutdown | Under Construction |
|---|---|---|---|---|---|---|---|---|---|---|---|---|---|
| FRANCE | 58 | 540,600.00 | 415,900.00 | 76.93% | 8 | | 2 | | | 60 | 1 | 12 | 1 |
| SLOVAKIA | 4 | 25,382.00 | 14,420.33 | 56.81% | | | | | | 8 | 1 | 3 | 2 |
| HUNGARY | 4 | 27,575.77 | 14,777.73 | 53.59% | | | | | | 4 | | | |
| UKRAINE | 15 | 168,284.00 | 83,122.77 | 49.39% | | | | | | 17 | 4 | 4 | 2 |
| BELGIUM | 7 | 67,556.95 | 32,093.75 | 47.51% | | | | | | 8 | | 1 | |
| SWEDEN | 10 | 150,163.00 | 62,270.04 | 41.47% | | 3 | | | 1 | 9 | | 3 | |
| SWITZ1ERLAND | 5 | 69,633.00 | 26,370.00 | 37.87% | | 2 | | | | 3 | 1 | 1 | |
| SLOVENIA | 1 | 16,269.00 | 6,060.82 | 37.25% | | | | | | 1 | | | |
| CZECH REPUBLIC | 6 | 80,045.10 | 28,636.78 | 35.78% | | | | | | 6 | | | |
| FINLAND | 4 | 65,378.00 | 22,654.00 | 34.65% | | 2 | | | | 3 | | | |
| BULGARIA | 2 | 47,221.00 | 15,866.65 | 33.60% | | | | | | 6 | | 4 | |
| ARMENIA | 1 | 7,389.00 | 2,266.00 | 30.67% | | | | | | 2 | | 1 | |
| KOREA, REPUBLIC OF | 24 | 490,372.00 | 149,165.00 | 30.42% | | | | | 4 | 24 | | | 4 |
| SPAIN | 7 | 268,400.00 | 54,832.00 | 20.43% | 1 | 2 | | | | 7 | | 3 | |
| UNITED STATES OF AMERICA | 99 | 4,092,935.00 | 797,067.00 | 19.47% | | 45 | 1 | 2 | 1 | 86 | 2 | 33 | 5 |
| RUSSIA | 34 | 910,400.00 | 169,048.55 | 18.57% | | | 2 | | | 28 | 18 | 5 | 9 |
| ROMANIA | 2 | 58,147.00 | 10,753.68 | 18.49% | | | | | 2 | | | | |
| UNITED KINGDOM | 16 | 337,030.00 | 57,918.48 | 17.18% | 41 | | 2 | | | 1 | 1 | 29 | |
| CANADA | 19 | 600,544.47 | 100,921.10 | 16.80% | | | | | 24 | | 1 | | 6 |
| GERMANY | 9 | 579,090.00 | 91,783.70 | 15.85% | | 11 | 1 | 2 | 1 | 20 | 1 | 27 | |
| SOUTH AFRICA | 2 | 237,841.00 | 14,748.56 | 6.20% | | | | | | 2 | | | |
| MEXICO | 2 | 165,066.00 | 9,311.61 | 5.64% | | 2 | | | | | | | |
| PAKISTAN | 3 | 106,268.15 | 4,609.69 | 4.34% | | | | | 1 | 4 | | | 2 |
| ARGENTINA | 3 | 129,747.63 | 5,257.70 | 4.05% | | | | | 3 | 1 | | | 1 |
| NETHERLANDS | 1 | 97,968.00 | 3,873.51 | 3.95% | | 1 | | | | 1 | | 1 | |
| INDIA | 21 | 941,413.96 | 33,231.88 | 3.53% | | 2 | 1 | | 22 | 2 | | | 6 |
| BRAZIL | 2 | 537,925.88 | 15,385.00 | 2.86% | | | | | | 3 | | | 1 |
| CHINA | 27 | 5,463,800.00 | 130,580.00 | 2.39% | | | 1 | 1 | 2 | 47 | | | 24 |
| IRAN, ISLAMIC REPUBLIC OF | 1 | 273,548.90 | 4,140.00 | 1.51% | | | | | | 1 | | | |
| JAPAN | 43 | 795,936.00 | 0.00 | 0.00% | | 35 | | | | 24 | 2 | 16 | 2 |
| Total | 438 | 17,351,930.81 | 2,377,066.33 | --- | 42 | 100 | 8 | 5 | 60 | 262 | 25 | | |

\* includes HWGCR, HWLWR, LWGR, WGR AND X

*DATA RECONSTRUCTED FROM IAEA – PRIS TABLES (RETRIEVED 07/01 – 07/11/2015 FROM HTTPS://WWW.IAEA.ORG/PRIS/WORLDSTATISTICS/OPERATIONALREACTORSBYCOUNTRY.ASPX*

There are aspects of the history of nuclear development that we must keep mind. As we've seen in the previous chapter on bombs vs. power there is an unfortunate historical coincidence of the beginnings of nuclear physics and the exigencies of WWII. Due to this, the Allies in WWII had already learned to enrich uranium. It is perhaps a little misleading to suggest that the ability to enrich uranium conditioned the infrastructure to use enriched uranium for power reactors but it seems quite logical and practical that it is one of the underlying factors.

# NUCLEAR ACCIDENTS

It is significant and unfortunate that most people do not discuss nuclear safety but are more than keen to discuss nuclear accidents regardless of the level of correct information they might have.  So, let us try to discuss what is officially and publicly known about the most significant events.

## INES - The International Nuclear and Radiological Event Scale

The IAEA has a Richter-like exponential scale for nuclear accidents that clearly and quickly identifies events from "incidents" to "accidents" on a number scale from the lowest 1 to the highest 7.

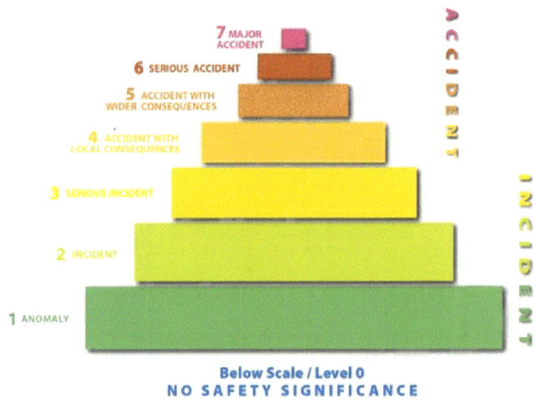

7 MAJOR ACCIDENT
6 SERIOUS ACCIDENT
5 ACCIDENT WITH WIDER CONSEQUENCES
4 ACCIDENT WITH LOCAL CONSEQUENCES
3 SERIOUS INCIDENT
2 INCIDENT
1 ANOMALY

ACCIDENT
INCIDENT

**Below Scale / Level 0**
**NO SAFETY SIGNIFICANCE**

*06/28/2015 FROM THE HOME PAGE OF THE IAEA INES WEBPAGE - HTTP://WWW-NS.IAEA.ORG/TECH-AREAS/EMERGENCY/INES.ASP*

*"THE PRIMARY PURPOSE OF INES IS TO FACILITATE COMMUNICATION AND UNDERSTANDING BETWEEN THE TECHNICAL COMMUNITY, THE MEDIA AND THE PUBLIC ON THE SAFETY SIGNIFICANCE OF EVENTS. THE AIM IS TO KEEP THE PUBLIC AS WELL AS NUCLEAR AUTHORITIES ACCURATELY INFORMED ON THE OCCURRENCE AND CONSEQUENCES OF REPORTED EVENTS." (HOME PAGE INES HTTP://WWW-NS.IAEA.ORG/TECH-REAS/EMERGENCY/INES.ASP. 06/28/2015)*

Based on an internationally accepted rating system the Table below summarizes **all** major accidents (from 4-7 on the INES scale).

**Nuclear "Events" / ACCIDENTS - INES Scale***
(most serious first)

| Location / Name | INES # | Date | Deaths | Exposures | Comment |
|---|---|---|---|---|---|
| Chernobyl, Ukraine | 7 | 4/26/1986 | 31 | still ? | Explosion & fire - widespread radiation release |
| Fukushima Daiichi, Japan | 7 | 3/11/2011 | 0 (up to 130) | ? | Tsunami - cooling water failure - partial meltdown |
| Kyshtym, Chelyabinsk Oblast, Russia | 6 | 9/29/1957 | 49-55 | ? | Plutonium processing for nuclear weapons - explosion - plume |
| Windscale, Great Britain | 5 | 10/10/1957 | 0 | 0? | Piles 1 & 2 were part of UK atomic bomb project - fire ... releases |
| Goiânia, Brazil | 5 | 9/13/1987 | 4 | 249 | Stolen radiotheraphy source from a local abandoned hosptial |
| Three Mile Island (TMI), USA | 5 | 3/28/1979 | 0 | 0 | Operator induced partial meltdown |
| Tokaimura | 4 | 9/30/1999 | 2 | 0 | Human error / safety violations |

* http://www-ns.iaea.org/tech-areas/emergency/ines.asp

Accepting the most liberal accounting of "nuclear" deaths, including theft of medical radiological sources (Brazil) the total of recorded deaths, at the high end is 222, from 1957 to post Fukushima (03/11/2011). For Fukushima, the projected maximum 130 deaths from increased cancer and other epidemiological markers has been added to the deaths making its contribution 58.56% of deaths regardless of the fact that there are no present casualties over 5 years post-incident.

Exposures to radiation present a different problem altogether. The Soviet Union was not very forthcoming with Chernobyl exposure information but that situation has ameliorated considerably[189] since the symbolic end of the cold war – the fall of the Berlin Wall on November 9, 1989. However, it is unlikely that data for the, then secret, Kyshtym, plutonium processing facility for weapons, was ever properly recorded or would ever be made available.

The 1957 Windscale fire and radioactive release (again from a weapons facility) in the United Kingdom got much more attention and study and even the most conservative estimates of exposure were revised upward in 2011[190] to 240 cancers (many thyroid cancers from radio-iodine).

The Three Mile Island (TMI) accident was a partial nuclear meltdown which occurred in Unit 2 of the two Three Mile Island nuclear reactors in Dauphin County, Pennsylvania, United States, on March 28, 1979. This was PWR designed by Babcock & Wilcox. It was the worst accident in U.S. nuclear power plant history. The problems were found to be related to an initial mechanical failure of a valve that was critically misinterpreted due to "Critical human factors and user interface engineering problems...

---

[189] A PBS Nature documentary "Radioactive Wolves" Full Episode Premiere date: October 19, 2011 (0:53:10) explored the Chernobyl exclusion zone in some detail.

[190] Story from BBC NEWS: http://news.bbc.co.uk/go/pr/fr/-/2/hi/science/nature/7030536.stm. Published: 2007/10/06 23:02:10 GMT. "Windscale fallout underestimated", Rebecca Morelle, Science reporter, BBC News.

revealed in the investigation of the reactor control system's user interface."[191]  It was abundantly clear from extensive reporting of the incident and from formal investigation that a major "Critical human factor" was inadequate training.

Despite the fact that TMI was the worst US nuclear accident and the overall perception (with the help of the media and at least one movie "Silkwood"[192]) that the incident was terribly dangerous **there were no deaths due to the incident and subsequent epidemiological studies showed no links to even a single post-accident cancer case.**[193,194,195,196]

**Radioactive Releases from Nuclear Accidents (TBq)\***

| Material | Windscale | Chernobyl | Fukushima Daiichi (atmospheric) | Three Mile Island |
|---|---|---|---|---|
| Iodine-131 | 740 | 1,760,000 | 130,000 | much less |
| Caesium-137 | 22 | 79,500 | 35,000 | much less |
| Xenon-133 | 12,000 | 6,500,000 | 17,000,000 | |
| Xenon-135 | | | | 25x Windscale |
| Strontium-90 | | 80,000 | | much less |
| Plutonium | | 6,100 | | |

* Tbq = terabecquerel is an SI unit of radioactivity. 1 Tbq = 27 curies (27 Ci)

Source: https://en.wikipedia.org/wiki/Windscale_fire.  Actual data from several internally referenced data sources.  Retrieved 07/18/2015.

---

[191] https://en.wikipedia.org/wiki/Three_Mile_Island_accident.  Retrieved 07/18/2015.

[192] "Silkwood" (1983) had nothing directly to do with TMI but has been definitely conflated with it.  The movie starred Meryl Streep, Kurt Russell and Cher.  http://www.imdb.com/title/tt0086312/.

[193] Maureen C. Hatch et al. (1990). "Cancer near the Three Mile Island Nuclear Plant: Radiation Emissions". American Journal of Epidemiology (Oxford Journals) **132**(3): 397–412.

[194] Hatch MC, Wallenstein S, Beyea J, Nieves JW, Susser M; Wallenstein; Beyea; Nieves; Susser (June 1991). "Cancer rates after the Three Mile Island nuclear accident and proximity of residence to the plant". American Journal of Public Health **81** (6): 719–724.

[195] RJ Levin (2008), "Incidence of thyroid cancer in residents surrounding the three-mile island nuclear facility", Laryngoscope **118** (4), pp. 618–628.

[196] http://www.scribd.com/doc/158526327/Settlement-of-Medical-Claims.

Although I am tempted to show a table of deaths due to coal mining and coal burning in fossil fuel plants, natural gas deaths and other such material and compare those numbers with the truly **miniscule numbers of nuclear fatalities** in the power industry – I will not.  **Suffice it to say mining and burning coal wins in all scales of disaster and loss, especially environmental damages!**

The Fukushima disaster presents an interesting example of several key problems that must be addressed and are touched upon in the "Musings" below.

# MUSINGS ON REACTOR DESIGN, OPERATION AND "ACCIDENTS"

**Is it possible to avoid human error completely? YES – in many, if not most circumstances,** especially if we remove individual humans from the loop when they cannot function properly. This has always seemed heretical to my past associates and even to some of my present ones. Teams of design people and troubleshooters can indeed devise a virtually foolproof system. The problem lies with human intervention in those carefully planned and designed systems, especially when the humans are under stress.

The trouble with human operators is that although they may be very good and intuitive in many things, they have severe limitations in others. After having personally been in a control room with reactor unit lights and alarms going off it is important to notice human reactions. Even the best trained operators will not be able to respond at a point when system alarms start to cascade. Human response times are limited and the ability to fully analyze complex problems or equipment failures always diminishes with reduced time.[197]

In general, most reactors, in-and-of-themselves, are well designed. Having said this there are many subsidiary facets of overall design that are poorly articulated or completely ignored. These elements are invariably external to basic engineering and fabrication and are more related to larger environmental factors or human variables.

A partial list of desired properties / behaviors is presented below for consideration:

- New Fission Reactors Should be "No-Meltdown"
- Geographically Specific Risk Factors must be addressed
- Operational Requirements
  - o Operator Machine Interface Optimization
  - o Internal Critical Self-sufficiency Assessment
  - o Guaranteed Safe Spin-Down

In considering the points above there will be a variety of recommendations that will not be popular amongst industry insiders as well as those outside but we must start to be inclusive in our considerations of design of any systems and one of the primary design elements that must be very seriously considered is the "human element".

---

[197] Buddaraju, Dileep. "PERFORMANCE OF CONTROL ROOM OPERATORS IN ALARM MANAGEMENT". A Thesis Submitted to the Graduate Faculty of the Louisiana State University and Agricultural and Mechanical College in Partial Fulfillment of the requirements for the degree of Master of Science in Engineering Science in The Interdepartmental Program in Engineering Science. May 2011. PDF available on-line.

# NEW FISSION REACTORS SHOULD BE "NO-MELTDOWN"

To reduce nuclear accident risk to the level of relatively banal, one simple and likely cost saving measure, would be to have an international agreement to build all new reactors in the general design concept of Canadian CANDU reactors. CANDUs are fueled with natural uranium and moderated with heavy water. The use of natural uranium virtually guarantees that the reactor will always go "sub-critical" – fail to produce enough neutrons to sustain a chain reaction – in the event of a major loss of coolant (LOC) accident. Using natural uranium removes any need for enrichment and any linkage to a nuclear weapons program.

The heavy water as the moderator is really the key. It absorbs very few neutrons (unlike light or regular water in PWRs and BWR's) and, when present, allows for chain reaction fission to occur. If there is a loss of coolant (moderator) accident the reactor immediately goes subcritical and chain reaction fission ceases. Of course, in the worst-case scenario, there will likely be damage to reactor components due to insufficient cooling but the possibility of meltdown is eradicated.

In terms of reducing costs one must take a larger view of the situation. If there is no necessity to do uranium enrichment in the first place, then costs will definitively decrease as will subsidiary exposure risks. A secondary benefit is that the reactor's fuel is not weapons grade and is not itself a potential target for terrorism *per se*.

PHWRs (specifically CANDU reactor) are really the intelligent choice for non-belligerent and humanely leaning regimes that are interested solely in nuclear power for its immense benefits to mankind. There is absolutely no rationale for running higher risk on any power reactor installation.

## GEOGRAPHICALLY SPECIFIC RISK FACTORS

The Fukushima Daiichi partial meltdown is an ideal situation to examine from the point of view of specific risk factors[198]. It is clear that the fundamental reactor design was very competent given that the reactor seemed to function completely as designed through the 9.0 $M_W$[199] Tōhoku earthquake that occurred at 14:46 on Friday, 11 March 2011[200]. The tsunami wave was sufficient to top the seawall and flooded the emergency diesel generators and cooling water pumps that were in a basement area known to be vulnerable to flooding[201]. The cooling water system was effectively destroyed for the reactor and a melt-down or partial melt-down was more-or-less inevitable. By all standards, the event was avoidable. Tepco knew that the sea wall was not high enough for a 100-year tsunami event but may have selected to take the chance that they would not see the 100-year tsunami event. There are a limited number of reactors in seismic zones that also have significant tsunami risk. These risks can be anticipated and mitigated with clear thinking and proper design … not cost cutting!

The entire volume of clean water needed to cool the reactor and to shut it down could total no more than a few hundred thousand gallons and a pumping system for a properly sequestered cooling water loop could have been incorporated into a revised reactor plan. If this inventory of water was held internally for emergency use and the appropriate secondary cooling put into place the entire situation could have been completely avoided.

Other emergency shutdown procedures might have included deliberate gadolinium "poisoning" ($^{157}$Gd that captures neutrons[202]) of the reactor to virtually instantly reduce neutron flux to avoid any runaway reaction. The rapid termination of the chain reaction fissioning is the best way to cool the core in an

---

[198] My intent here is to not reference an easily accessible and large number of articles on Fukushima. The details can be gleaned on a moment to moment basis for the crisis. The response was flawed and many measures questionably executed. My intention is to point to the meta-idea of fundamental design strategy.

[199] "The **moment magnitude scale** (abbreviated as **MMS**; denoted as **M$_W$** or **M**) is used by seismologists to measure the size of earthquakes in terms of the energy released.[1] The magnitude is based on the seismic moment of the earthquake, which is equal to the rigidity of the Earth multiplied by the average amount of slip on the fault and the size of the area that slipped.[2] The scale was developed in the 1970s to succeed the 1930s-era Richter magnitude scale (M$_L$). Even though the formulae are different, the new scale retains the familiar continuum of magnitude values defined by the older one. The MMS is now the scale used to estimate magnitudes for all modern large earthquakes by the United States Geological Survey.[3]" https://en.wikipedia.org/wiki/Moment_magnitude_scale. Retrieved 07/20/2015.

[200] https://en.wikipedia.org/wiki/Fukushima_Daiichi_nuclear_disaster. Retrieved 07/20/2015.

[201] Apparently, an in-house study regarding the dangers of flooding for the cooling system was ignored by Tepco officials. "TEPCO did not act on tsunami risk projected for nuclear plant". https://jagadees.wordpress.com/2012/02/13/tepco-did-not-act-on-tsunami-risk-projected-for-nuclear-plant/. Retrieved 07/20/2015.

[202] G. Leinweber,* D. P. Barry, M. J. Trbovich, J. A. Burke, N. J. Drindak, H. D. Knox, and R. V. Ballad. "Neutron Capture and Total Cross-Section Measurements and Resonance Parameters of Gadolinium". *NUCLEAR SCIENCE AND ENGINEERING*: **154**, 261–279 (2006).

emergency situation. Using a salt like gadolinium nitrate, even in seawater (as a last resort) used to cool the core would leave the gadolinium in place even if the water evaporated, thereby successfully capturing neutrons and averting the meltdown in a more efficient manner.

There are several references in the Fukishima literature that point to harried officials and inadequate emergency responses that are touched on in the next section.

## OPERATIONAL REQUIREMENTS

### OPERATOR MACHINE INTERFACE OPTIMIZATION

The problems with humans making complex decisions under stressful situations has been the topic of extensive psychological, sociological and anthropological study across a wide range of fields, problems and applications. Regardless of the scenario there is a critical point at which intelligent and measured decisions cannot be made by human operators. UNDER SUCH CONDITIONS (that can be easily quantified) HUMAN DECISION MUST BE SUPERCEDED BY MACHINE DECISIONS.

### INTERNAL CRITICAL SELF-SUFFICIENCY ASSESSMENTS

ALL MAJOR SYSTEMS must be critically self-sufficient in ALL SCENARIOS OF EXTREME POTENTIAL DAMAGE AND RISK. This means that they must have the ability to shut down internally regardless of external conditions. This requires autonomous control systems that are integrated on higher operational levels.

Is this costly? Yes. Is it done routinely? No. Is it possible to do? Yes. Why is it not done? Generally, $$$'s (and often fragile egos).

### GUARANTEED SAFE SPIN-DOWN

Only those who have had immediate experience with very high power applications and equipment have been made to understand the entire concept of "spin-down". In a system with large turbines and generators with considerable momentum … work … heat, often a considerable amount of time is required to shut the system down without damage to the immediate equipment or severe disruptions to grid power regulation and distribution. The term "spin-down" is from the actual spinning of massive turbines that are doing the energy transfer from whatever the primary heat production process is to the production of power.

Turning off any power production system is not at all like turning off a lightbulb. Doing it improperly (generally too rapidly) can be very damaging. That is why it is very important to couple the safe reactor shutdown to the safe spin-down time in the secondary circuit so that heat from the reactor can be bled off, thus reducing the risk of damages, overreactions and unnecessary leaks.

All major power users should turn their priorities to the building of many more safe reactors and **even toward international oversight of reactor design and siting. It may even be more radical to suggest but if we, collectively, are brave enough to do the right thing then we can go that extra step and select the safest one-design system possible (dare I again strongly suggest CANDU – PHWR's?).** This less-than-subtle suggestion has already been introduced in the requirement for no melt-down.

Even though I am not fond of bureaucracy I am explicitly suggesting that there be a global nuclear initiative and control. It would be my fondest hope that it would be more a technocracy than bureaucracy – an agency run by scientists and engineers who had the mandate, responsibility and authority to oversee operations and maintenance in all nuclear power facilities. In the present world, this seems intensely naïve but when a solution to a problem is clear that solution should be pursued regardless of the immediate perceptions and potential difficulties.

The primary resistance to this notion will undoubtedly be the politicians themselves who are the ones who would not step-up-to-the-plate to solve the climate crisis in the first place. As with most things, the people who maintain 100% control are the least likely to voluntarily relinquish that control.

All of the other asides and presentation on fusion, fission, bombs vs. power, and perception versus reality are essentially part of the first step in sorting out the global climate crisis. We must think and act more rationally, more clearly and, in a manner that decouples perception from reality. We cannot afford the luxury or lunacy of "alternative facts" – just the real, often uncomfortable facts.

# OTHER SOLUTIONS ...

## SOLUTIONS THAT CAN BEGIN IMMEDIATELY

Public relations and education campaigns to improve public opinion of the necessity to rely significantly more on nuclear energy – both fission and fusion. This was already introduced previously but is repeated here because of its fundamental importance within the sphere of things we can do and should be doing to improve global climate.

Funnel uncommitted energy research money into fusion experimental projects or simply cancel them and put the money into the best candidates. Attempt to drastically accelerate the international "Iter" project OR supercede it. It may seem heretical that the very project that I would personally love to ring in as a success is the very project that I fear is doomed for two fundamental reasons:

1. The multinational consortium is not bound by any hard commitments deadlines.
2. The present nominal deadline is 35 years! That is more than half my life, at least five times longer than it took to produce an atomic bomb, FAR TOO LONG TO MAKE ANY DIFFERENCE FOR THE PRESENT AND VERY REAL CLIMATE PROBLEM.

## PASSIVE-AGGRESSIVE SOLUTIONS

These solutions are the most difficult for a straightforward scientist to formulate because the primary reason for most of them are socio-economic and political variables. These variables (people and their attitudes) are often difficult to quantitate or predict.

## SYTEMATIC AND RELENTLESS RE-EDUCATION AND LOBBY CAMPAIGN

The primary elements of the campaign would be:

- establish an accurate historic perspective and education program focused on fundamental energy transfer considerations and energy yields of materials
- establish REAL environmental and risk factors for fossil fuels, hydroelectric and nuclear power
- MASSIVELY increased effort to get widespread public support for accelerated fusion research and development leading to an effective fusion reactor. The best research teams in fusion should be included up-front in this effort.
- increase support for fission reactors with reduced construction schedules (not laxer environmental assessments but faster and more effective vetting).

## ALTER NATIONAL TAX STRUCTURE(S) TO ACCELERATE ENERGY INFRASTRUCTURE CHANGE

Fossil fuel producers should be assessed a carbon tax that will be used directly to fund renewable energy projects and heighten a nuclear transition. The present notion of buying and selling carbon credits seemingly does nothing to encourage an overall change in base use patterns – all it does is divert focus from the fundamentals.

## PURSUE GLOBAL ECONOMIC PRESSURE TO REDUCE FOSSIL FUEL BURNING

Although this is a non-focused and general recommendation it is critical that the major fossil fuel users show a real example by cutting back and offering economic incentives (best done by favorable trade agreements) for other nations to follow suit. Given the international economic competitiveness the reduction of fossil fuel usage would operationally mean substantial increases in nuclear and renewable energy. Although many nations are theoretically pursing these goals the real measure of success is not intent ...it is the relative increase in renewable energy versus present consumption that should be the target. As has been pointed out in prior sections, the fact that a nation may be the leader in renewable energy investment is TOTALLY MEANINGLESS IF THEIR ACTUAL PERCENTAGE OF RENEWABLE ENERGY IS NOT INCREASING IN EXCESS OF THEIR FOSSIL FUEL CONSUMPTION.

## AGGRESSIVE SOLUTIONS

What the title of this book promises is aggressive solutions. These are ranked as the least aggressive first and build up rapidly as the perceived and real problems mount.

Operational logic requires that the solutions are initially invoked at the national level and, much more rarely, at the International level. Despite this general rule of thumb, we are considering a global problem that requires a global solution. **Perhaps the best way to start this discussion is to present historical data of significant rapid global climate changes.**

Dramatic and non-cyclical climatic changes have definitely occurred in the past and the most dramatic was the comet impact that resulted in the destruction of the dinosaurs approximately 66 million years ago. The global winter that killed off the dinosaurs was short lived but sufficient to cause an immense change in the ecosystems of the earth. There is a considerable amount of evidence and general scientific acceptance of what was formerly known as the KT-extinction or KT-boundary effect, it is now referred to as the K-Pg boundary.[203]  Regardless of unfortunate naming, the event that occurred was the impact of a meteorite or asteroid slightly larger than the area of Manhattan Island (22.82 mi²) at the tip of the Yucatan Peninsula.

> "Such a large impact would have had approximately the energy of 100 trillion tons of TNT, or about 2 million times greater than the most powerful thermonuclear bomb ever tested."[204]

The very dramatic Chicxulub impactor[205] (asteroid /comet) event was truly global as indicated by the very thin and distinct iridium layer verified for that geological stratum (date) in many locations around the globe.  There is general scientific agreement as to what occurred but there is ongoing research as to the specific origin of the asteroid.  The Alverez hypothesis[206] that it was a chondritic meteorite is well accepted based on the global iridium concentrations within the specific stratum of interest and other supporting observations.

In addition to the very interesting and catastrophic impact (for dinosaurs … but positive for mammals) of Chicxulub there are many other smaller, much more recent, and much more reliably observed events that can be used to scale any intervention that we might be forced to entertain as global warming becomes more serious.

The eruption of Mt. Pinatubo, Luzon, Philippines on June 15, 1991, was the second largest eruption of the 20th century[207] and provided a very good test case for injection of particulate into the stratosphere and its global effects.  Its near-equatorial location was ideal for rapid world-wide transport of the $SO_2$ aerosol and particulate plume in the stratosphere.

---

[203] "The Cretaceous–Paleogene (K–Pg) boundary, formerly known as the Cretaceous–Tertiary (K–T) boundary,[a] is a geological signature, usually a thin band of rock.  K, the first letter of the German word *Kreide* (chalk), is the traditional abbreviation for the Cretaceous Period and Pg is the abbreviation for the Paleogene Period." https://en.wikipedia.org/wiki/Cretaceous%E2%80%93Paleogene_boundary.  Retrieved 12/05/2016.

[204] Ibid.

[205] "The Chicxulub impactor (/ˈtʃiːkʃəluːb/ cheek-shə-loob), also known as the K/Pg impactor and (more speculatively) as the Chicxulub asteroid, was an asteroid or comet at least 10 kilometres (6 miles) in diameter which created the Chicxulub crater. It impacted a few miles from the present-day town of Chicxulub in Mexico, after which the impactor and its crater are named. Because the estimated date of the object's impact and the Cretaceous–Paleogene boundary (K–Pg boundary) coincide, there is a scientific consensus that its impact was the Cretaceous–Paleogene extinction event which caused the demise of the planet's non-avian dinosaurs and other species.[3][4]
The impactor's crater is more than 180 km (110 miles) in diameter,[5] making it the third largest known impact crater on Earth." https://en.wikipedia.org/wiki/Chicxulub_impactor.  Retrieved 12/06/2016.

[206] Alvarez, LW; Alvarez, W; Asaro, F & Michel, HV (1980). "Extraterrestrial cause for the Cretaceous–Tertiary extinction". Science. 208 (4448): 1095–1108. Bibcode:1980Sci...208.1095A.

[207] The largest was Novarupta, in the Aleutian Range in Alaska.  It had a VEI=6 and was active from June 6 to October of 1912.  See https://en.wikipedia.org/wiki/Novarupta.

Pinatubo was an event that was really the first large scale volcanic eruption to be studied with a satellite platform with both optical and infra-red capabilities. The Stratospheric Aerosol and Gas Experiment II (SAGE II) was flying aboard NASA's Earth Radiation Budget Satellite (ERBS). The four views below provide a useful starting point to be able to correlate the Volcanic Eruption Index (VEI)[208] with airborne aerosols, dust, other gases, optical clarity measurements of the atmosphere and, most importantly, with global temperature changes.

SAGE II 1020 nm Optical Depth

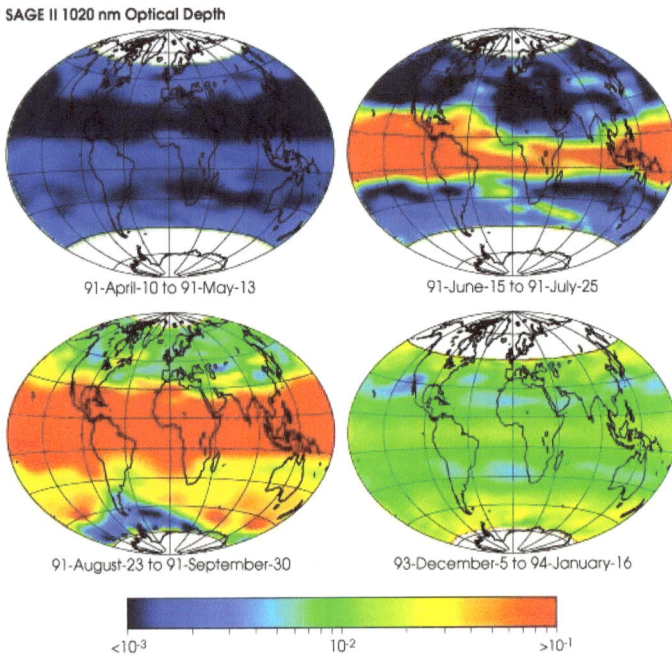

91-April-10 to 91-May-13

91-June-15 to 91-July-25

91-August-23 to 91-September-30

93-December-5 to 94-January-16

$<10^{-3}$         $10^{-2}$         $>10^{-1}$

The images above were acquired by the Stratospheric Aerosol and Gas Experiment II (SAGE II) flying aboard NASA's Earth Radiation Budget Satellite (ERBS). The false-color images represent aerosol optical depth in the stratosphere during four different time spans, ranging from before the June 1991 Pinatubo eruption to two years after the event. Red pixels show the highest values, while dark blue shows the lowest values, which are normally observed in the stratosphere. Notice how the volcanic plume gradually spreads across virtually the entire globe, hence the global-scale impact on climate.

http://earthobservatory.nasa.gov/IOTD/view.php?id=1510. Retrieved 12/06/2016.

---

[208] The VEI was devised by Chris Newhall of the United States Geological Survey and Stephen Self at the University of Hawaii in 1982. It is a relative scale from 0 (non-explosive) to 8 ("mega-colossal") - the largest eruptions in history. https://en.wikipedia.org/wiki/Volcanic_Explosivity_Index. Retrieved 03/16/2017.

The updated Sage III platform was designed to function from the International Space Station and was launched by SpaceX on a Falcon 9 booster from the Kennedy Space Center, Florida on February 19, 2017. It will provide much higher resolution of aerosol and gas data.

## Mauna Loa Apparent Transmission

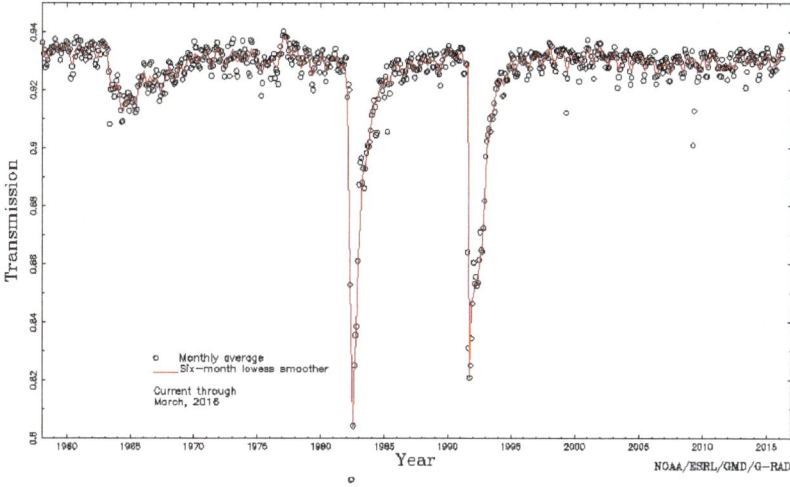

The "apparent" transmission, or transmission ratio (Ellis & Pueschel, *Science*, 1971), is derived from broadband (0.3 to 2.8um) direct solar irradiance observations at the Mauna Loa Observatory (19.533 ° N, 155.578 ° W, elev. 3.4 km) in Hawaii. Data are for clear-sky mornings between solar elevations of 11.3 and 30 degrees. The plotted points are monthly averages and the plotted curve results from a 6 month lowess statistical smoother.

http://www.esrl.noaa.gov/gmd/grad/mloapt.html. Retrieved 12/06/2016.

The first peak in the diagram above corresponds to the eruption of El Chichón (March-April 1982). The solar transmission ratio for the Mt. Pinatubo eruption (second peak at mid-1991) does not recover to the baseline until late 1995 to early 1996. This is a strong indication of the expected length of the atmospheric change caused by periodic injections of materials into the stratosphere (3-5 years is a pretty good bet).

Pinatubo is important because of its impressive scale; it is really the type of scale one would be looking for to have any real effect on global climate. There are many other recent eruptions where there has been stratospheric injection of material and considerable downwind dispersion. Studying these events carefully will be useful in setting overall parameters for any climate interventions. These studies would also be useful for carrying out more localized climate change[209] such as those limited to certain latitudes.

The most powerful eruption in recent history was on April 10, 1815. Mount Tambora (on the island of Sumbawa in present day Indonesia) erupted with an estimate VEI = 7. The 1815 eruption caused a long lasting, extreme climate event in 1816 known as the "year without a summer." A type of volcanic winter affected most of the northern hemisphere.

> "…the smaller particles spewed by the volcano were light enough to spread through the atmosphere over the following months and had a worldwide effect on climate. They made their way into the stratosphere, where they could distribute around the world more easily. Earth's average global temperature dropped three degrees Celsius."[210]

Unfortunately, there was no optical or other analytical data but the temperature anomalies were easily tracked.

Alan Robock wrote a very good paper on "Volcanic Eruptions and Climate"[211] that clearly indicates the strong and somewhat predictable effects of larger volcanic events on global climate. There is some difficulty in scaling the climate effects to smaller events than Tambora, El Chichon, and Pinatubo.

On December 29, 2008, Koryaksky erupted with a 6,000 m (20,000 ft) plume of ash, the first major eruption in 3,500 years.[212] This would have been of sufficient altitude to enter the lower stratosphere but there is seemingly no visible effect in the apparent atmospheric transmission (see previous diagram). This is quite understandable as Koryaksky is at the tip of the Kamchatka Peninsula of Russia at coordinates 53°19'15"N 158°42'45"E whereas the observation station is on Mauna Loa Observatory on Hawaii at 19.533 ° N, 155.578 ° W. To cause any observable change in atmospheric transmission the eruption would have had to have been both significant and long-lasting enough to permit the global mixing of particulates and gasses within the stratosphere.

Smaller scale and shorter eruptions that are at upper latitudes will likely be more easily detected by the Sage III installation on the ISS. Monitoring these and correlating with any atmospheric transmissivity changes and temperature measurements would be a critical step in moving towards measured and correlated climatic interventions.

---

[209] What I particularly mean here is that if the detonation sites are very far north or south (> 50 degrees latitude) the effects of stratospheric injection will likely not be global. They will indeed follow the upper level jets but mixing to lower latitudes can often be seasonally anticipated.

[210] UCAR (NCAR) on-line article, "Mount Tambora and the Year Without a Summer". https://scied.ucar.edu/shortcontent/mount-tambora-and-year-without-summer. Retrieved 03/02/2017.

[211] Robock, Alan. "Volcanic Eruptions and Climate". **Reviews of Geophysics**, 38, 2 / May 2000 pages 191–219.

[212] "Russian volcano erupts". BBC NEWS. December 30, 2008. Retrieved 03/14/2017.

## STRATOSPHERIC INJECTION OF AEROSOLS

Having considered the effects of natural volcanism and spectacular events like comet impacts it is very clear that injection of particulate and aerosol materials into the stratosphere can have significant effects on local and global climate.

This has not escaped the scientific community and there have been several papers on the subject of manmade injections of aerosols into the stratosphere[213,214]

This "solution" is more of a treatment than a cure. The cure is to not create the $CO_2$ excesses in the first place.

Launching dedicated aircraft on missions to seed the stratosphere with aerosols would be a very stupid and wasteful idea. Every "dedicated" flight would just add more $CO_2$ to the global burden. If aircraft delivery of stratospheric aerosol is necessary or, more importantly, would provide a significant climatic effect, then the most sensible approach would be to find ways to equip commercial aircraft to distribute small quantities of aerosols as part of their normal flights. Due to the sheer volume of commercial air traffic, significant quantities could be delivered without adding significantly to the $CO_2$ production.

The real questions in such an approach are how much will be enough aerosol to have a significant effect and whether this type of mitigation will have a secondary negative effect of empowering the major polluters to continue their use patterns more-or-less unchecked simply because there is a simple "fix". There are always negative effects of "messing with nature" but if we are pushed to going this far there should be an exit strategy.

---

[213] Hulme, Mike. "Climate change: Climate engineering through stratospheric aerosol injection." Progress in Physical Geography. 2012 36: 694 originally published online 9 August 2012.

[214] Walia, Arjun. "TOP CLIMATE SCIENTIST WARNS AGAINST INJECTING STRATOSPHERIC PARTICLES INTO THE ATMOSPHERE". January 12, 2015. http://www.collective-evolution.com/2015/01/12/top-climate-scientist-warns-against-injecting-stratospheric-particles-into-the-atmosphere/. Retriever 02/21/2017.

# SYNTHETIC VOLCANIC WINTER – A HIGHLY AGGRESSIVE SOLUTION

Volcanos are a large potential contributor to climate change. Certainly, we have no control over incoming comets like the one that extinguished dinosaurs so we must focus on more "local" resources.

The best approach to altering climate is always the minimal one! Using volcanoes to provide an effective means of large-scale stratospheric injection of aerosols and particulates that will lower global average temperature is the best first choice for human-climate intervention. The next obvious question is how can we manipulate volcanos to do anything that we have in mind for climate control?

That answer is brutally and surprisingly self-evident. The extent to which we can manipulate any natural forces is directly proportional to the energy we can direct to whatever process we wish to "enhance". In this case, it would be ideal to stimulate volcanic activity of sufficient power to result in stratospheric injection of sulfate aerosols and particulates that will measurably cause non-local decreases in atmospheric temperature.

Due to the scale of volcanos and the scale of eruptive event that should occur, the smallest explosive event that would facilitate stratospheric injection of ash and aerosols would have to be, at very minimum, hundreds to thousands of tons of TNT. Use of conventional explosives of this mass and under the demanding terrain and seismic activity of the active volcano would be a virtual impossibility and exceptionally dangerous for personnel. For this inherently practical reason alone the only viable alternative would be to use tactical nuclear weapons.

If this solution were under consideration there would be a minimal basis-set of conditions or selection criteria. The research materials for this section are strongly tied to classified information from nuclear test detonations. Given that these were very highly controlled and extensively studied it is reasonably safe to say that the data will be very useful in determining the selection of significant variables:

- parameters for maximal stratospheric injection of aerosols and particulate matter
- local seismic disruption
- ability to effectively and continuously monitor plume conditions.

The primary elements would be site selection:

- isolation from downwind population and contamination sensitive areas
- as high an elevation as possible (logically, if the target is getting material into the stratosphere with the minimum necessary quantum yield device then, greater elevation is favorable). Of course, there are no volcanoes even close to the lower stratosphere, but the but the higher the altitude the less the tropospheric mixing can interfere with injection.

Even considering the very basic conditions above, it does not take a great deal of imagination to begin to discuss potential sites in the world.

One candidate site with a secondary positive aspect would be Mt. Erebus[215] on Ross Island, Antarctica. It is the second largest volcano on Antarctica and has one of the longest periods of activity. This would be a good site to couple the natural volcanic output with the explosive power of a relatively small tactical nuclear device. These are basically artillery shells with low yield warheads that were designed to be used in close proximity to ones' own troops. There was significant development of these warheads during the cold war and they were considered a strong deterrent to the Soviet Union's purported threat to the European allies of the United States (mostly NATO countries). They range in size from very small (72 tons of TNT equivalent yield or .072 megatons for the W48 155-millimeter nuclear artillery shell) to very large shells, or missile warheads, exceeding the yields of Hiroshima and Nagasaki.[216]

Climate change may indeed become climate crisis and require desperation solutions. However, even these must be entertained in as thoughtful, controlled, and phased a manner is possible. That is a good reason to start with small-scale experiments and very careful monitoring and prediction in order to step up the pace and magnitude as required.

Another positive point for using Mt. Erebus is that the circulating stratosphere flow will limit the extent of the spread of materials and will provide a more reliable temperature correlation to aerosol burden.

In the northern hemisphere, sites are somewhat more limited in that the relative land area to ocean is higher and the population density at the higher northern latitudes is greater than the higher southern latitudes. Another factor for consideration is potential disruption of air travel. There have been a number of volcanic incidents where airliners have encountered severe engine problems[217] due to volcanic ash. One of the more hazardous places is in the Iceland air corridor due to the common routing of intercontinental flights.

Another possible location for a tactical nuclear device coupled to a potential volcanic event would be in the Kamchatka Peninsula of Russia. There are approximately 160 volcanos and 29 are presently active.[218] The zone may be too far south (57° 0' 0" N, 160° 0' 0" E) to be a good test area because the prevailing northern hemisphere winds in that zone are west to east and the downwind areas are

[215] https://en.wikipedia.org/wiki/Mount_Erebus. Retrieved 03/17/2017.
[216] https://en.wikipedia.org/wiki/Tactical_nuclear_weapon. Retrieved 03/17/2017.
[217] "Why Can't Planes Fly Through Volcanic Ash? NASA Found Out the Hard Way", by Clay Dillow. April 19, 2010. Popular Science on-line. http://www.popsci.com/science/article/2010-04/why-cant-planes-fly-through-volcanic-ash-because-nasa-tried-once. Retrieved 03/17/2017.
[218] https://en.wikipedia.org/wiki/Volcanoes_of_Kamchatka. Retrieved 03/18/2017.

populated areas of Alaska and Canada. However, if injections were truly stratospheric this would be a somewhat irrelevant objection. There is a "fall-out" zone of particulate and gaseous emission in the immediate vicinity of the volcano at lower altitudes that is carried by local tropospheric winds (downwind).

This following suggested site of a "low-yield nuclear / volcanic excitation" *might be safely attributed to the new Trump administration* ... perhaps using Mount Paektu, in North Korea as a test target[219], for an ALCM with a W80[220] thermonuclear warhead. It is highly unlikely that the North Koreans could detect it and the positive repercussions would be some (at least local) climate change with the added benefit that the North Korean government would have their hands full dealing with the eruption aftereffects.[221]

---

[219] Excuse me but I cannot exclude my dark sense of humor along with the more serious side of selecting potential sites.

[220] https://en.wikipedia.org/wiki/W80_(nuclear_warhead). Variable capability from 5 to 15 kilotons. Retrieved 03/23/2017.

[221] I hope that this is taken as an unnecessary display of my personal macabre sense of humor, but given Trump and his immediate advisors, I would not be surprised if they jumped on it as a great idea!

# SYNTHETIC NUCLEAR WINTER – A DESPARATION SOLUTION

This is the most extreme solution to be offered up sooner than most sane people would have expected even a few years ago. One of the very unfortunate potential causes of perhaps bringing this solution in closer temporal proximity is the disturbing United States energy policy that is being articulated by the new Donald Trump administration[222]. Planned reductions of automobile air quality standards and increasing use of coal will have some very negative global effects on climate that will certainly accelerate the temperature increases.

Synthetic or Nuclear winter is very simply that – a period of global cooling caused by the CAREFULLY CALCULATED and SAFELY DETONATED NUCLEAR WARHEADS that will inject as much particulate and aerosol material into the stratosphere as necessary to lower global temperature systematically.

This entire notion is simply a continuation of the prior section of using volcanos as the site of origin of the explosions. However, the implicit scale is much larger and the sites could be carefully selected to maximize atmospheric effects and safety. Also, the site selection is not limited to volcanos.

WE SHOULD NEVER WANT TO HAVE TO GO TO A MEASURE THIS DRASTIC but ... given the present global environmental and political situation we may very well have to seriously consider such desperate and damaging "solutions" to the climate change problem that could have been prevented with some foresight!

We often seem to forget that being proactive is a much better position to be in rather than being reactive.

---

[222] Rick Perry was only confirmed as Energy Secretary on March 2, 2017. This passage was initially written on 03/17/2017 so there is no discernable policy in place at the moment but it is highly likely to be consistent with Trump campaign promises and the former Texas governor's support for the oil and gas industry in general.

# IN SUMMARY

**THERE IS NO DOUBT** that China, the United States, India, Russia and Japan are the culprits but there is a critical component of any potential solution that is missing – ownership and responsibility for the problem. Everyone must be highly suspect of the agreements on climate that any of these countries have seemingly come to, given the scientifically verifiable failure of these countries in controlling $CO_2$.

Science and technology are the fundamental engines of human advancement – **not politics or religion.** Therefore, science and a scientific approach must lead the way to a better cleaner world. We must begin to be rational about problems we share as human beings and attempt to develop a framework for problem resolution and development of intelligent and scientifically credible public policy.

## WHY IS CLIMATE CHANGE IMPORTANT FOR EVERYONE?

1. Increase in sea level will cause enormous expense and effort to relocate populations in coastal areas.
2. Large expanses of ice like the Ross Ice Shelf in Antarctica are disappearing[223]. This is habitat for krill – a major source of food for marine animals higher up in the food change.
3. Arctic changes in ice formation are already threatening the existence of polar bears.
4. The Great Barrier Reef of Australia is being decimated at an unprecedented rate. This is a major loss of marine habitat and biodiversity that the planet requires for overall "Health".
5. Increase in desertification – this is critically important in two major ways:
    a. loss of the vital $CO_2$ conversion capability of vegetation (another contribution to warming).
    b. loss of arable land.
    c. extensive habitat loss and species endangerment.
6. Potentially large yield losses for crops that are not drought tolerant. This will push marginal agricultural areas to the brink of famine.
7. Increased demand for irrigation water and the negative additional energy use that it implies.

This is only a partial list and it cannot account for the incredible interwoven dependency of all interacting living things with their native environments. When intrinsic native environments shift rapidly many creatures and plants are not equipped to adapt for survival. The Chicxulub impactor of 65 million years ago that effectively destroyed all the major dinosaurs is a good example the effects of excessively rapid climate change.

---

[223] While doing the final edits before posting to CreateSpace, yet another disturbing piece of news on CNN about Antarctic ice conditions appeared on the Internet – A Delaware-sized piece of ice is calving from the Larsen C. ice shelf (07/25/2017).

## SOLUTIONS

A fundamental re-education of people regarding energy production and its effects. This will have to include the undeniable fact that fossil fuel burning is not environmentally sound policy.

An aggressive pro-nuclear campaign with fusion being the absolute top priority. The building of new fission reactors as safely and quickly as possible is necessary to buffer the already alarming rise in global temperatures.

- New Fission Reactors Should be "No-Meltdown"
- Geographically Specific Risk Factors must be explicitly accounted for
- Operational Requirements
    - Operator Machine Interface Optimization
    - Internal Critical Self-sufficiency Assessment
    - Guaranteed Safe Spin-Down

Political realignment by outside electoral pressure or the by the very slight possibility that enlightened leadership will emerge internally. Most politicians are married to the short-term maintenance of power and control of perception of people – not on reality or the necessary actions to effect real solutions to problems. This is particularly the case when the solutions are costly and, by that measure alone, unpopular.

Scientists (especially government scientists) should be expected to speak clearly and unequivocally to politicians. Those who know must be held responsible for the consequences of allowing critical decisions to be made by politicians who have no real knowledge about climate change or science in general.

People of all faiths must come together internationally to wield their political power to enforce meaningful environments and energy policy. Perhaps a global, Vatican-instigated, initiative on the need for Planetary Energy Management could be the starting point for a new dialogue and vision of the future.

We must prepare ourselves with the best possible research base to initiate the more drastic solutions suggested should they become necessary:

- Deliberate activation of volcanic eruptions – to increase stratospheric aerosols and slow warming
- Tactical nuclear explosions to increase stratospheric aerosols (same reasons as above).

# CONCLUSION

*I dearly hope that the "conclusion" is not the perpetuation of the inane behaviors that have shaped the last one hundred fifty years – exploitation, environmental destruction, feigned horror, belated correction and retrenchment to the same modes of thinking and behavior.  **THERE ARE OTHER WAYS!***

*Many of the things that I have suggested may seem to be quasi-impossible but we must all refuse to engage in that negative stream of self-rationalization.  We have all contributed to a rich, diverse and productive world.  If we can do that we can muster the courage, resources, and real commitment to make the necessary changes that can make our world sustainable and provide as stable a global environment as nature will permit.*

*If this is "GUSHY GREEN RHETORIC" so be it!*

# ACKNOWLEDGEMENTS

Dr. Ann McMillan – editor and important contributor who added some depth and dimension to several of the simple points that I was trying to make. Most importantly, she moderated some of my personal tendency to rant. She is a rare government scientist who has had the integrity to step forward and write the foreword.

Celeste Bryant – ever patient editor who is fundamentally concerned that real people can read and understand the text. She also encouraged me to restart the writing when I had basically given up in early 2016.

Dr. Ajeet Jon Saxon for editorial assistance at the meta level. His overall worldly wisdom is important to have onboard when one is personally distressed and motivated to prove something even when virtually nobody really cares. As a former engineer turned economist he would have been a better author than I but was far too intelligent to take on that task and even smarter to maintain some level of credible deniability. Probably not sufficient deniability to protect him from associative criticism.

Dr. Bill (William) Leith, Program Coordinator for the Earthquake Hazards Program, Global Seismographic Network Program, and National Geomagnetism Program. Bill was very helpful in steering my questions on volcanism and monitoring to the active people in the field with the qualified research answers.

Dr. Kayla Iacovino, Volcanologist, at Arizona State University, who was kind enough to try to answer my rather outré questions about volcanoes, not laugh, and give very balanced answers.

To the pond scum scientists [you know damn well who you are!] in government and academic positions who haven take no real action when they know that they should and who would prefer to make a show of marching rather than acting. You make writing the more vituperative segments of the text relatively easy and fun!

To the politicians who have given up any real chance to make a difference and who simply tell people what they want to hear to get elected for one more useless term. Congratulations you arrogant, useless, spineless boils on the ass of humanity!

To the people and organizations out there struggling to make a real difference in a world of "bread and circuses". I salute you. You do something that I am very bad at.

# SELECTED ONLINE RESOURCES – CAVEAT EMPTOR!

Given the qualification in the heading to this section it should be exceptionally clear that when using any Internet resources, one's "Bullshit Sensors" must be tuned to maximum sensitivity.

The American Chemical Society (ACS)

http://www.acs.org/

> There is an accessible and instructive Climate Science Toolkit

http://www.acs.org/content/acs/en/climatescience.html

Ballotpedia

"Ballotpedia is the online encyclopedia of American politics and elections. Our goal is to inform people about politics by providing accurate and objective information about politics at all levels of government. We are firmly committed to neutrality; here's why. Ballotpedia's articles are 100 percent written by our professional staff… Ballotpedia currently has over 220,000 encyclopedic articles … Ballotpedia is sponsored by the Lucy Burns Institute, a nonpartisan and nonprofit organization headquartered in Middleton, Wisconsin."

> Retrieved 12/17/2016 – ellipses indicate omissions of background material that is easily accessed on their website.  Founded on 2007.

https://ballotpedia.org

CleanTechnica

"*CleanTechnica* is the #1 cleantech-focused website in the US and the world according to Compete.com and Quantcast.com."

> Retrieved from the website "About" page 02/22/2017.

https://cleantechnica.com

The Climate Council (Australia)

"After thousands of Australians chipped in to Australia's biggest crowd-funding campaign, the abolished Climate Commission has relaunched as the new, independent Climate Council.

**We exist to provide independent, authoritative climate change information to the Australian public. Why? Because our response to climate change should be based on the best science available.**"

> Retrieved from the website under the tab "About Us" on 11/18/2016

https://www.climatecouncil.org.au/


Climate Nexus

"Climate Nexus is a strategic communications organization dedicated to changing the conversation on climate and clean energy solutions in the United States."

http://climatenexus.org/

> This seems to be much more of a PR organization than anything else. It is not possible to detect a large component of technical expertise and judgement.


Index Mundi

**"About IndexMundi**

IndexMundi is a data portal that gathers facts and statistics from multiple sources and turns them into easy to use visuals.

Our mission is to turn raw data from all over the world into useful information for a global audience. We capture statistics that are scattered or otherwise hidden and present them via user-friendly maps, charts, and tables which allow visitors to understand complex information at a glance."

http://www.indexmundi.com/


IPCC (Intergovernmental Panel on Climate Change)

https://www.ipcc.ch/index.htm

Ipsos

"An independent company, managed and controlled by research professionals, Ipsos offers deep specialist expertise, market leading methodologies and the ability to execute comprehensive and reliable global research programs through offices in 87 countries."

Abstracted from the website "About Ipsos" tab 02/16/2017.

http://www.ipsos-na.com


International Atomic Energy Agency (IAEA)

https://www.iaea.org/

IAEA PRIS (Power Reactor Information System)

https://www.iaea.org/PRIS/


Iter (The Iter Tokamak)

"ITER" ("The Way" in Latin) is one of the most ambitious energy projects in the world today.

In southern France, 35 nations are collaborating to build the world's largest tokamak, a magnetic fusion device that has been designed to prove the feasibility of fusion as a large-scale and carbon-free source of energy based on the same principle that powers our Sun and stars.

The experimental campaign that will be carried out at ITER is crucial to advancing fusion science and preparing the way for the fusion power plants of tomorrow.

ITER will be the first fusion device to produce net energy. ITER will be the first fusion device to maintain fusion for long periods of time. And ITER will be the first fusion device to test the integrated technologies, materials, and physics regimes necessary for the commercial production of fusion-based electricity.

Thousands of engineers and scientists have contributed to the design of ITER since the idea for an international joint experiment in fusion was first launched in 1985. The ITER Members—**China, the European Union, India, Japan, Korea, Russia and the United States**—are now engaged in a 35-year collaboration to build and operate the ITER experimental device, and together bring fusion to the point where a demonstration fusion reactor can be designed."

Abstracted from the "About" page 02/17/2017.

https://www.iter.org/

This is a long-term project that has great goals but the international collaboration and co-funding of operations may become a bigger problem than the ambitious program. Also, the time delays to first plasma (2025) and to operational goals (2035) *may be far too long*.

The International Energy Agency

https://www.iea.org

This is an autonomous agency, has existed since 1974 and is based in France:

International Energy Agency

31-35 rue de la Fédération

75739 Paris Cedex 15 France

It has (as of 11/29/2016) 29 member countries.

https://www.iea.org/countries/membercountries/

NASA – Earth Observatory

http://earthobservatory.nasa.gov/Features/WorldOfChange/decadaltemp.php

National Institute for Fusion Science - Japan

http://www.lhd.nifs.ac.jp/en/

The Large Helical Device (LHD) information:

http://www.lhd.nifs.ac.jp/en/home/lhd.html

Oak Ridge National Laboratory

https://www.ornl.gov/

"Oak Ridge National Laboratory is the largest US Department of Energy science and energy laboratory, conducting basic and applied research to deliver transformative solutions to compelling problems in energy and security".  Taken from page 1 of 2 – downloaded Fact Sheet 02/14/2017.

A useful resource for news on fusion research is available form periodic newsletters available at:

http://web.ornl.gov/info/stelnews/stelnews.html

The Planetary Science Institute

"The Planetary Science Institute is a private, nonprofit 501(c)(3) corporation dedicated to Solar System exploration. It is headquartered in Tucson, Arizona, where it was founded in 1972. In 2016, established a second office in Lakewood, CO, near Denver."

https://www.psi.edu/

Skeptical Science

http://www.skepticalscience.com/


United States Department of Energy (DOE)

http://www.energy.gov/


United States Energy Information Administration (EIA)

www.eia.gov

>This is a very useful information source that can be manipulated in several ways to generate relevant data. As is always the case one must be very careful in formulating proper queries and in citations of specific data sets or they will not be comparable.


United States Geological Survey (USGS)

https://www.usgs.gov/

>Their website is dense and very informative on a very broad range of environmental issues.


The World Bank

"The World Bank is like a cooperative, made up of 189 member countries. These member countries, or shareholders, are represented by a Board of Governors, who are the ultimate policymakers at the World Bank. Generally, the governors are member countries' ministers of finance or ministers of development. They meet once a year at the Annual Meetings of the Boards of Governors of the World Bank Group and the International Monetary Fund."

**Headquarters**
1818 H Street, NW Washington, DC 20433 USA (202) 473-1000

>http://www.worldbank.org/    Main site
>
>http://data.worldbank.org/    Tab on main site page "DATA"

World Economic Forum

"It was established in 1971 as a not-for-profit foundation and is headquartered in Geneva, Switzerland. It is independent, impartial and not tied to any special interests."  Retrieved from the "ABOUT" tab on the website 12/08/2016.

https://www.weforum.org/

# APPENDIX

## IAEA REACTOR TYPE GLOSSARY (ABSTRACTED)

The IAEA has a convenient glossary[224] from which the short list of fission reactor types (with acronyms shown below) was abstracted. Please note that squared brackets [ ] were added by the author for clarity[225]. It is important to be aware of reactor type since they are indeed not all the same and safety as well as nuclear proliferation issues are involved.

Advanced Gas-cooled Reactor (AGR) – There is now a second generation of British gas-cooled reactors, using graphite as the neutron moderator and carbon dioxide as a coolant.

Boiling Water Reactor (BWR) - Boiling light water cooled and moderated reactor. In a BWR, the reactor core heats water, which turns to steam and then drives a steam turbine.

Fast Breeder Reactor (FBR) - Fast neutron reactors use fast neutrons to cause fission in their fuel. They do not have a neutron moderator, and use less-moderating coolants.

Gas Cooled Reactor (GCR) - A gas-cooled reactor is a nuclear reactor that uses graphite as a neutron moderator and carbon dioxide (helium can also be used) as coolant.

[HTGR = High Temperature Gas-cooled Reactor – variant of type]

[HWGCR = Heavy Water Gas Cooled Reactor. The only others of this type are listed as EL-4 (MONTS D'ARREE) in France and BOHUNICE A1 in Slovakia, both permanently shut down).]

[HWLWR = Heavy Water / Light Water reactor. The only one listed is Fugen ATR in Japan and it is permanently shut down).]

Light Water Cooled Graphite Moderated Reactor (LWGR) - Also known as RBMK. Uses light water for cooling and graphite for moderation, it is also possible to use natural uranium for fuel.

Pressurized Heavy Water Reactor (PHWR) - A pressurized heavy water reactor is a nuclear power reactor, commonly using unenriched natural uranium as its fuel that uses heavy water (deuterium oxide $D_2O$) as its coolant and moderator.

[Steam-generating heavy water reactor (SGHWR) – This is virtually the same as a CANDU PHWR. The only type designation was for Winfrith in the United Kingdom and it permanently shut down having successful reached its design life.]

Pressurized Water Reactor (PWR) - Pressurized light water moderated and cooled reactor. In a PWR, the reactor core heats water, which does not boil. This hot water then exchanges heat with a lower pressure water system, which turns to steam and drives the turbine.

---

[224] https://www.iaea.org/PRIS/Glossary.aspx. Retrieved 07/05/2015.
[225] When using the IAEA – PRIS database online the glossary one will find that it does not cover some the reactor types listed under individual countries so these descriptions were added in the squared brackets.

[X = Who the h____ knows?  ... It likely means experimental.  The only instances of this reactor type entry are for the United States – 2 reactors of the 99 total; Hallam [226]and Piqua[227] now permanently shut down).]

[226] The Hallam Nuclear Generating Station in Nebraska was the site of an experimental graphite-moderated sodium cooled reactor.  It was decommissioned in 1969.  For more details see www.wikipedia.org/wiki/Hallam_Nuclear_Generating_Station.

[227] The Piqua Nuclear Power Facility was a nuclear power plant which operated just outside the southern city limits of Piqua, Ohio in the United States.  The plant contained a 45.5-megawatt (thermal) organically cooled and moderated nuclear reactor (terphenyl, a biphenyl like oil).  For more details see www.wikipedia.org/wiki/Piqua_Nuclear_Generating_Station.